Simulation of Fresh Concrete Flow

RILEM STATE-OF-THE-ART REPORTS
Volume 15

RILEM, The International Union of Laboratories and Experts in Construction Materials, Systems and Structures, founded in 1947, is a non-governmental scientific association whose goal is to contribute to progress in the construction sciences, techniques and industries, essentially by means of the communication it fosters between research and practice. RILEM's focus is on construction materials and their use in building and civil engineering structures, covering all phases of the building process from manufacture to use and recycling of materials. More information on RILEM and its previous publications can be found on www.RILEM.net.

The RILEM State-of-the-Art Reports (STAR) are produced by the Technical Committees. They represent one of the most important outputs that RILEM generates - high level scientific and engineering reports that provide cutting edge knowledge in a given field. The work of the TCs is one of RILEM's key functions.

Members of a TC are experts in their field and give their time freely to share their expertise. As a result, the broader scientific community benefits greatly from RILEM's activities.

RILEM's stated objective is to disseminate this information as widely as possible to the scientific community. RILEM therefore considers the STAR reports of its TCs as of highest importance, and encourages their publication whenever possible.

The information in this and similar reports is mostly pre-normative in the sense that it provides the underlying scientific fundamentals on which standards and codes of practice are based. Without such a solid scientific basis, construction practice will be less than efficient or economical.

It is RILEM's hope that this information will be of wide use to the scientific community.

For further volumes:
http://www.springer.com/series/8780

Nicolas Roussel · Annika Gram
Editors

Simulation of Fresh Concrete Flow

State-of-the-Art Report of the RILEM Technical Committee 222-SCF

Springer

Editors
Nicolas Roussel
Université Paris Est
IFSTTAR
Marne la Vallée Cedex 2
France

Annika Gram
CBI
Stockholm
Sweden

ISSN 2213-204X ISSN 2213-2031 (electronic)
ISBN 978-94-017-8883-0 ISBN 978-94-017-8884-7 (eBook)
DOI 10.1007/978-94-017-8884-7
Springer Dordrecht Heidelberg New York London

Library of Congress Control Number: 2014934297

Printed on acid-free paper

Springer is part of Springer Science+Business Media (www.springer.com)

RILEM Publications

RILEM Publications are presented in 6 collections, corresponding to the 5 clusters of active RILEM Technical Committees, sorted by fields of expertise, and a 6th multi-thematic collection dedicated to journals and compendiums:

A. Mechanical Performance and Fracture
B. Test Methods, Materials Characterization and Processing
C. Service Life and Design
D. Durability and Deterioration Mechanisms
E. Bitumen, Masonry and Timber
F. Journals and Compendiums

Each publication is assigned to one of the following series: reports (REP), proceedings (PRO), compendiums (COMP) and journals. The former CD-ROM series is now included in one of these series.

Each publication is available in at least one of the three following editions: print (PR), CD or DVD-ROM (CD), or online (OL).

Online editions are available through our web site, at http://www.rilem.net

The RILEM DVD-ROM, gathering several thousands of online articles, is also published and updated each year (internal publication, circulation restricted to RILEM Benefactor Members).

The following list is presenting our global offer, sorted by series.

RILEM Proceedings

PRO 1: Durability of High Performance Concrete (ISBN: 2-912143-03-9); *Ed. H. Sommer*

PRO 2: Chloride Penetration into Concrete (ISBN: 2-912143-00-04); *Eds. L.-O. Nilsson and J.-P. Ollivier*

PRO 3: Evaluation and Strengthening of Existing Masonry Structures (ISBN: 2-912143-02-0); *Eds. L. Binda and C. Modena*

PRO 4: Concrete: From Material to Structure (ISBN: 2-912143-04-7); *Eds. J.-P. Bournazel and Y. Malier*

PRO 5: The Role of Admixtures in High Performance Concrete (ISBN: 2-912143- 05-5); *Eds. J.G. Cabrera and R. Rivera-Villarreal*

PRO 6: High Performance Fiber Reinforced Cement Composites - HPFRCC 3 (ISBN: 2-912143-06-3); *Eds. H.W. Reinhardt and A.E. Naaman*

PRO 7: 1st International RILEM Symposium on Self-Compacting Concrete (ISBN: 2-912143-09-8); *Eds. Å. Skarendahl and Ö. Petersson*

PRO 8: International RILEM Symposium on Timber Engineering (ISBN: 2-912143-10-1); *Ed. L. Boström*

PRO 9: 2nd International RILEM Symposium on Adhesion between Polymers and Concrete ISAP '99 (ISBN: 2-912143-11-X); *Eds. Y. Ohama and M. Puterman*

PRO 10: 3rd International RILEM Symposium on Durability of Building and Construction Sealants (ISBN: 2-912143-13-6); *Eds. A.T. Wolf*

PRO 11: 4th International RILEM Conference on Reflective Cracking in Pavements (ISBN: 2-912143-14-4); *Eds. A.O. Abd El Halim, D.A. Taylor and El H. H. Mohamed*

PRO 12: International RILEM Workshop on Historic Mortars: Characteristics and Tests (ISBN: 2-912143-15-2); *Eds. P. Bartos, C. Groot and J.J. Hughes*

PRO 13: 2nd International RILEM Symposium on Hydration and Setting (ISBN: 2-912143-16-0); *Ed. A. Nonat*

PRO 14: Integrated Life-Cycle Design of Materials and Structures - ILCDES 2000 (ISBN: 951-758-408-3); (ISSN: 0356-9403); *Ed. S. Sarja*

PRO 15: Fifth RILEM Symposium on Fibre-Reinforced Concretes (FRC) - BEFIB'2000 (ISBN: 2-912143-18-7); *Eds. P. Rossi and G. Chanvillard*

PRO 16: Life Prediction and Management of Concrete Structures (ISBN: 2-912143-19-5); *Ed. D. Naus*

PRO 17: Shrinkage of Concrete – Shrinkage 2000 (ISBN: 2-912143-20-9); *Eds. V. Baroghel-Bouny and P.-C. Aïtcin*

PRO 18: Measurement and Interpretation of the On-Site Corrosion Rate (ISBN: 2-912143-21-7); *Eds. C. Andrade, C. Alonso, J. Fullea, J. Polimon and J. Rodriguez*

PRO 19: Testing and Modelling the Chloride Ingress into Concrete (ISBN: 2-912143-22-5); *Eds. C. Andrade and J. Kropp*

PRO 20: 1st International RILEM Workshop on Microbial Impacts on Building Materials (CD 02) (e-ISBN 978-2-35158-013-4); *Ed. M. Ribas Silva*

PRO 21: International RILEM Symposium on Connections between Steel and Concrete (ISBN: 2-912143-25-X); *Ed. R. Eligehausen*

PRO 22: International RILEM Symposium on Joints in Timber Structures (ISBN: 2-912143-28-4); *Eds. S. Aicher and H.-W. Reinhardt*

PRO 23: International RILEM Conference on Early Age Cracking in Cementitious Systems (ISBN: 2-912143-29-2); *Eds. K. Kovler and A. Bentur*

PRO 24: 2nd International RILEM Workshop on Frost Resistance of Concrete (ISBN: 2-912143-30-6); *Eds. M.J. Setzer, R. Auberg and H.-J. Keck*

PRO 25: International RILEM Workshop on Frost Damage in Concrete (ISBN: 2-912143-31-4); *Eds. D.J. Janssen, M.J. Setzer and M.B. Snyder*

PRO 26: International RILEM Workshop on On-Site Control and Evaluation of Masonry Structures (ISBN: 2-912143-34-9); *Eds. L. Binda and R.C. de Vekey*

PRO 27: International RILEM Symposium on Building Joint Sealants (CD03); *Ed. A.T. Wolf*

PRO 28: 6th International RILEM Symposium on Performance Testing and Evaluation of Bituminous Materials - PTEBM'03 (ISBN: 2-912143-35-7; e-ISBN: 978-2-912143-77-8); *Ed. M.N. Partl*

PRO 29: 2nd International RILEM Workshop on Life Prediction and Ageing Management of Concrete Structures (ISBN: 2-912143-36-5); *Ed. D.J. Naus*

PRO 30: 4th International RILEM Workshop on High Performance Fiber Reinforced Cement Composites - HPFRCC 4 (ISBN: 2-912143-37-3); *Eds. A.E. Naaman and H.W. Reinhardt*

PRO 31: International RILEM Workshop on Test and Design Methods for Steel Fibre Reinforced Concrete: Background and Experiences (ISBN: 2-912143-38-1); *Eds. B. Schnütgen and L. Vandewalle*

PRO 32: International Conference on Advances in Concrete and Structures 2 vol. (ISBN (set): 2-912143-41-1); *Eds. Ying-shu Yuan, Surendra P. Shah and Heng-lin Lü*

PRO 33: 3rd International Symposium on Self-Compacting Concrete (ISBN: 2-912143-42-X); *Eds. Ó. Wallevik and I. Níelsson*

PRO 34: International RILEM Conference on Microbial Impact on Building Materials (ISBN: 2-912143-43-8); *Ed. M. Ribas Silva*

PRO 35: International RILEM TC 186-ISA on Internal Sulfate Attack and Delayed Ettringite Formation (ISBN: 2-912143-44-6); *Eds. K. Scrivener and J. Skalny*

PRO 36: International RILEM Symposium on Concrete Science and Engineering – A Tribute to Arnon Bentur (ISBN: 2-912143-46-2); *Eds. K. Kovler, J. Marchand, S. Mindess and J. Weiss*

PRO 37: 5th International RILEM Conference on Cracking in Pavements – Mitigation, Risk Assessment and Prevention (ISBN: 2-912143-47-0); *Eds. C. Petit, I. Al-Qadi and A. Millien*

PRO 38: 3rd International RILEM Workshop on Testing and Modelling the Chloride Ingress into Concrete (ISBN: 2-912143-48-9); *Eds. C. Andrade and J. Kropp*

PRO 39: 6th International RILEM Symposium on Fibre-Reinforced Concretes - BEFIB 2004 (ISBN: 2-912143-51-9); *Eds. M. Di Prisco, R. Felicetti and G.A. Plizzari*

PRO 40: International RILEM Conference on the Use of Recycled Materials in Buildings and Structures (ISBN: 2-912143-52-7); *Eds. E. Vázquez, Ch. F. Hendriks and G.M.T. Janssen*

PRO 41: RILEM International Symposium on Environment-Conscious Materials and Systems for Sustainable Development (ISBN: 2-912143-55-1); *Eds. N. Kashino and Y. Ohama*

PRO 42: SCC'2005 - China: 1st International Symposium on Design, Performance and Use of Self-Consolidating Concrete (ISBN: 2-912143-61-6); *Eds. Zhiwu Yu, Caijun Shi, Kamal Henri Khayat and Youjun Xie*

PRO 43: International RILEM Workshop on Bonded Concrete Overlays (e-ISBN: 2-912143-83-7); *Eds. J.L. Granju and J. Silfwerbrand*

PRO 44: 2nd International RILEM Workshop on Microbial Impacts on Building Materials (CD11) (e-ISBN: 2-912143-84-5); *Ed. M. Ribas Silva*

PRO 45: 2nd International Symposium on Nanotechnology in Construction, Bilbao (ISBN: 2-912143-87-X); *Eds. Peter J.M. Bartos, Yolanda de Miguel and Antonio Porro*

PRO 46: ConcreteLife'06 - International RILEM-JCI Seminar on Concrete Durability and Service Life Planning: Curing, Crack Control, Performance in Harsh Environments (ISBN: 2-912143-89-6); *Ed. K. Kovler*

PRO 47: International RILEM Workshop on Performance Based Evaluation and Indicators for Concrete Durability (ISBN: 978-2-912143-95-2); *Eds. V. Baroghel-Bouny, C. Andrade, R. Torrent and K. Scrivener*

PRO 48: 1st International RILEM Symposium on Advances in Concrete through Science and Engineering (e-ISBN: 2-912143-92-6); *Eds. J. Weiss, K. Kovler, J. Marchand, and S. Mindess*

PRO 49: International RILEM Workshop on High Performance Fiber Reinforced Cementitious Composites in Structural Applications (ISBN: 2-912143-93-4); *Eds. G. Fischer and V.C. Li*

PRO 50: 1st International RILEM Symposium on Textile Reinforced Concrete (ISBN: 2-912143-97-7); *Eds. Josef Hegger, Wolfgang Brameshuber and Norbert Will*

PRO 51: 2nd International Symposium on Advances in Concrete through Science and Engineering (ISBN: 2-35158-003-6; e-ISBN: 2-35158-002-8); *Eds. J. Marchand, B. Bissonnette, R. Gagné, M. Jolin and F. Paradis*

PRO 52: Volume Changes of Hardening Concrete: Testing and Mitigation (ISBN: 2-35158-004-4; e-ISBN: 2-35158-005-2); *Eds. O.M. Jensen, P. Lura and K. Kovler*

PRO 53: High Performance Fiber Reinforced Cement Composites - HPFRCC5 (ISBN: 978-2-35158-046-2); *Eds. H.W. Reinhardt and A.E. Naaman*

PRO 54: 5th International RILEM Symposium on Self-Compacting Concrete (ISBN: 978-2-35158-047-9); *Eds. G. De Schutter and V. Boel*

PRO 55: International RILEM Symposium Photocatalysis, Environment and Construction Materials (ISBN: 978-2-35158-056-1); *Eds. P. Baglioni and L. Cassar*

PRO56: International RILEM Workshop on Integral Service Life Modelling of Concrete Structures (ISBN 978-2-35158-058-5); *Eds. R.M. Ferreira, J. Gulikers and C. Andrade*

PRO57: RILEM Workshop on Performance of cement-based materials in aggressive aqueous environments (e-ISBN: 978-2-35158-059-2); *Ed. N. De Belie*

PRO58: International RILEM Symposium on Concrete Modelling - CONMOD'08 (ISBN: 978-2-35158-060-8); *Eds. E. Schlangen and G. De Schutter*

PRO 59: International RILEM Conference on On Site Assessment of Concrete, Masonry and Timber Structures - SACoMaTiS 2008 (ISBN set: 978-2-35158-061-5); *Eds. L. Binda, M. di Prisco and R. Felicetti*

PRO 60: Seventh RILEM International Symposium on Fibre Reinforced Concrete: Design and Applications - BEFIB 2008 (ISBN: 978-2-35158-064-6); *Ed. R. Gettu*

PRO 61: 1st International Conference on Microstructure Related Durability of Cementitious Composites 2 vol., (ISBN: 978-2-35158-065-3); *Eds. W. Sun, K. van Breugel, C. Miao, G. Ye and H. Chen*

PRO 62: NSF/RILEM Workshop: In-situ Evaluation of Historic Wood and Masonry Structures (e-ISBN: 978-2-35158-068-4); *Eds. B. Kasal, R. Anthony and M. Drdácký*

PRO 63: Concrete in Aggressive Aqueous Environments: Performance, Testing and Modelling, 2 vol., (ISBN: 978-2-35158-071-4); *Eds. M.G. Alexander and A. Bertron*

PRO 64: Long Term Performance of Cementitious Barriers and Reinforced Concrete in Nuclear Power Plants and Waste Management - NUCPERF 2009 (ISBN: 978-2-35158-072-1); *Eds. V. L'Hostis, R. Gens, C. Gallé*

PRO 65: Design Performance and Use of Self-consolidating Concrete - SCC'2009 (ISBN: 978-2-35158-073-8); *Eds. C. Shi, Z. Yu, K.H. Khayat and P. Yan*

PRO 66: 2nd International RILEM Workshop on Concrete Durability and Service Life Planning - ConcreteLife'09 (ISBN: 978-2-35158-074-5); *Ed. K. Kovler*

PRO 67: Repairs Mortars for Historic Masonry (e-ISBN: 978-2-35158-083-7); *Ed. C. Groot*

PRO 68: Proceedings of the 3rd International RILEM Symposium on 'Rheology of Cement Suspensions such as Fresh Concrete (ISBN 978-2-35158-091-2); *Eds. O.H. Wallevik, S. Kubens and S. Oesterheld*

PRO 69: 3rd International PhD Student Workshop on 'Modelling the Durability of Reinforced Concrete (ISBN: 978-2-35158-095-0); *Eds. R.M. Ferreira, J. Gulikers and C. Andrade*

PRO 70: 2nd International Conference on 'Service Life Design for Infrastructure' (ISBN set: 978-2-35158-096-7, e-ISBN: 978-2-35158-097-4); *Ed. K. van Breugel, G. Ye and Y. Yuan*

PRO 71: Advances in Civil Engineering Materials - The 50-year Teaching Anniversary of Prof. Sun Wei' (ISBN: 978-2-35158-098-1; e-ISBN: 978-2-35158-099-8); *Eds. C. Miao, G. Ye, and H. Chen*

PRO 72: First International Conference on 'Advances in Chemically-Activated Materials – CAM'2010' (2010), 264 pp, ISBN: 978-2-35158-101-8; e-ISBN: 978-2-35158-115-5, *Eds. Caijun Shi and Xiaodong Shen*

PRO 73: 2nd International Conference on 'Waste Engineering and Management - ICWEM 2010' (2010), 894 pp, ISBN: 978-2-35158-102-5; e-ISBN: 978-2-35158-103-2, *Eds. J.Zh. Xiao, Y. Zhang, M.S. Cheung and R. Chu*

PRO 74: International RILEM Conference on 'Use of Superabsorsorbent Polymers and Other New Addditives in Concrete' (2010) 374 pp., ISBN: 978-2-35158-104-9; e-ISBN: 978-2-35158-105-6; *Eds. O.M. Jensen, M.T. Hasholt, and S. Laustsen*

PRO 75: International Conference on 'Material Science - 2nd ICTRC - Textile Reinforced Concrete - Theme 1' (2010) 436 pp., ISBN: 978-2-35158-106-3; e-ISBN: 978-2-35158-107-0; Ed. W. Brameshuber

PRO 76: International Conference on 'Material Science - HetMat - Modelling of Heterogeneous Materials - Theme 2' (2010) 255 pp., ISBN: 978-2-35158-108-7; e-ISBN: 978-2-35158-109-4; Ed. W. Brameshuber

PRO 77: International Conference on 'Material Science - AdIPoC - Additions Improving Properties of Concrete - Theme 3' (2010) 459 pp., ISBN: 978-2-35158-110-0; e-ISBN: 978-2-35158-111-7; *Ed. W. Brameshuber*

PRO 78: 2nd Historic Mortars Conference and RILEM TC 203-RHM Final Workshop – HMC2010 (2010) 1416 pp., e-ISBN: 978-2-35158-112-4; *Eds. J. Válek, C. Groot, and J.J. Hughes*

PRO 79: International RILEM Conference on Advances in Construction Materials Through Science and Engineering (2011) 213 pp., e-ISBN: 978-2-35158-117-9; *Eds. Christopher Leung and K.T. Wan*

PRO 80: 2nd International RILEM Conference on Concrete Spalling due to Fire Exposure (2011) 453 pp., ISBN: 978-2-35158-118-6, e-ISBN: 978-2-35158-119-3; *Eds. E.A.B. Koenders and F. Dehn*

PRO 81: 2nd International RILEM Conference on Strain Hardening Cementitious Composites (SHCC2-Rio) (2011) 451 pp., ISBN: 978-2-35158-120-9, e-ISBN: 978-2-35158-121-6; *Eds. R.D. Toledo Filho, F.A. Silva, E.A.B. Koenders and E.M.R. Fairbairn*

PRO 82: 2nd International RILEM Conference on Progress of Recycling in the Built Environment (2011) 507 pp., e-ISBN: 978-2-35158-122-3; *Eds. V.M. John, E. Vazquez, S.C. Angulo and C. Ulsen*

PRO 83: 2nd International Conference on Microstructural-related Durability of Cementitious Composites (2012) 250 pp., ISBN: 978-2-35158-129-2; e-ISBN: 978-2-35158-123-0; *Eds. G. Ye, K. van Breugel, W. Sun and C. Miao*

PRO 85: RILEM-JCI International Workshop on Crack Control of Mass Concrete and Related issues concerning Early-Age of Concrete Structures – ConCrack 3 – Control of Cracking in Concrete Structures 3 (2012) 237 pp., ISBN: 978-2-35158-125-4; e-ISBN: 978-2-35158-126-1; *Eds. F. Toutlemonde and J.-M. Torrenti*

PRO 86: International Symposium on Life Cycle Assessment and Construction (2012) 414 pp., ISBN: 978-2-35158-127-8, e-ISBN: 978-2-35158-128-5; *Eds. A. Ventura and C. de la Roche*

PRO 87 draft: UHPFRC 2013 – RILEM-fib-AFGC International Symposium on Ultra-High Performance Fibre-Reinforced Concrete (2013), ISBN: 978-2-35158-130-8, e-ISBN: 978-2-35158-131-5; *Eds. F. Toutlemonde*

PRO 88: 8th RILEM International Symposium on Fibre Reinforced Concrete (2012) 344 pp., ISBN: 978-2-35158-132-2, e-ISBN: 978-2-35158-133-9; *Eds. Joaquim A.O. Barros*

PRO 90: 7th RILEM International Conference on Self-Compacting Concrete and of the 1st RILEM International Conference on Rheology and Processing of Construction Materials (2013) 396 pp, ISBN: 978-2-35158-137-7, e-ISBN: 978-2-35158-138-4, *Eds. Nicolas Roussel and Hela Bessaies-Bey*

RILEM Reports

Report 19: Considerations for Use in Managing the Aging of Nuclear Power Plant Concrete Structures (ISBN: 2-912143-07-1); *Ed. D.J. Naus*
Report 20: Engineering and Transport Properties of the Interfacial Transition Zone in Cementitious Composites (ISBN: 2-912143-08-X); *Eds. M.G. Alexander, G. Arliguie, G. Ballivy, A. Bentur and J. Marchand*
Report 21: Durability of Building Sealants (ISBN: 2-912143-12-8); *Ed. A.T. Wolf*
Report 22: Sustainable Raw Materials - Construction and Demolition Waste (ISBN: 2-912143-17-9); *Eds. C.F. Hendriks and H.S. Pietersen*
Report 23: Self-Compacting Concrete state-of-the-art report (ISBN: 2-912143-23-3); *Eds. Å. Skarendahl and Ö. Petersson*
Report 24: Workability and Rheology of Fresh Concrete: Compendium of Tests (ISBN: 2-912143-32-2); *Eds. P.J.M. Bartos, M. Sonebi and A.K. Tamimi*
Report 25: Early Age Cracking in Cementitious Systems (ISBN: 2-912143-33-0); *Ed. A. Bentur*
Report 26: Towards Sustainable Roofing (Joint Committee CIB/RILEM) (CD 07) (e-ISBN 978-2-912143-65-5); *Eds. Thomas W. Hutchinson and Keith Roberts*
Report 27: Condition Assessment of Roofs (Joint Committee CIB/RILEM) (CD 08) (e-ISBN 978-2-912143-66-2); *Ed. CIB W 83/RILEM TC166-RMS*
Report 28: Final report of RILEM TC 167-COM 'Characterisation of Old Mortars with Respect to Their Repair (ISBN: 978-2-912143-56-3); *Eds. C. Groot, G. Ashall and J. Hughes*
Report 29: Pavement Performance Prediction and Evaluation (PPPE): Interlaboratory Tests (e-ISBN: 2-912143-68-3); *Eds. M. Partl and H. Piber*
Report 30: Final Report of RILEM TC 198-URM 'Use of Recycled Materials' (ISBN: 2-912143-82-9; e-ISBN: 2-912143-69-1); *Eds. Ch. F. Hendriks, G.M.T. Janssen and E. Vázquez*
Report 31: Final Report of RILEM TC 185-ATC 'Advanced testing of cement-based materials during setting and hardening' (ISBN: 2-912143-81-0; e-ISBN: 2-912143-70-5); *Eds. H.W. Reinhardt and C.U. Grosse*
Report 32: Probabilistic Assessment of Existing Structures. A JCSS publication (ISBN 2-912143-24-1); *Ed. D. Diamantidis*
Report 33: State-of-the-Art Report of RILEM Technical Committee TC 184-IFE 'Industrial Floors' (ISBN 2-35158-006-0); *Ed. P. Seidler*
Report 34: Report of RILEM Technical Committee TC 147-FMB 'Fracture mechanics applications to anchorage and bond' Tension of Reinforced Concrete Prisms – Round Robin Analysis and Tests on Bond (e-ISBN 2-912143-91-8); *Eds. L. Elfgren and K. Noghabai*
Report 35: Final Report of RILEM Technical Committee TC 188-CSC 'Casting of Self Compacting Concrete' (ISBN 2-35158-001-X; e-ISBN: 2-912143-98-5); *Eds. Å. Skarendahl and P. Billberg*
Report 36: State-of-the-Art Report of RILEM Technical Committee TC 201-TRC 'Textile Reinforced Concrete' (ISBN 2-912143-99-3); *Ed. W. Brameshuber*

Report 37: State-of-the-Art Report of RILEM Technical Committee TC 192-ECM 'Environment-conscious construction materials and systems' (ISBN: 978-2-35158-053-0); *Eds. N. Kashino, D. Van Gemert and K. Imamoto*
Report 38: State-of-the-Art Report of RILEM Technical Committee TC 205-DSC 'Durability of Self-Compacting Concrete' (ISBN: 978-2-35158-048-6); *Eds. G. De Schutter and K. Audenaert*
Report 39: Final Report of RILEM Technical Committee TC 187-SOC 'Experimental determination of the stress-crack opening curve for concrete in tension' (ISBN 978-2-35158-049-3); *Ed. J. Planas*
Report 40: State-of-the-Art Report of RILEM Technical Committee TC 189-NEC 'Non-Destructive Evaluation of the Penetrability and Thickness of the Concrete Cover' (ISBN 978-2-35158-054-7); *Eds. R. Torrent and L. Fernández Luco*
Report 41: State-of-the-Art Report of RILEM Technical Committee TC 196-ICC 'Internal Curing of Concrete' (ISBN 978-2-35158-009-7); *Eds. K. Kovler and O. M. Jensen*
Report 42: 'Acoustic Emission and Related Non-destructive Evaluation Techniques for Crack Detection and Damage Evaluation in Concrete' - Final Report of RILEM Technical Committee 212-ACD (e-ISBN: 978-2-35158-100-1); *Ed. M. Ohtsu*

Introduction

Nicolas Roussel

Université Paris Est, Ifsttar, France

What are the final objectives of the extensive research, which has been carried out in the last fifty years on rheology of fresh concretes?

A researcher answer could be: "the understanding of the correlation between mix design and rheological properties" or "the ability to correctly measure and quantify the rheological properties of concrete". These are of course points of great interest but a practitioner would however probably answer: "the ability to predict whether or not a given concrete will correctly fill a given formwork".

An analogy with the state of knowledge in the hardened concrete research field can be made: a lot of work has been indeed carried out in order to understand the correlation between mechanical properties and mix design and many tests have been developed in order to measure these mechanical properties (mechanical strength and delayed deformations for instance) but, on the other hand, many developments were also carried out in the field of structural engineering in order to correlate the needed properties of the concrete to be cast with the structure to be built. This last step has been missing for years in the rheology field. Only recently, researchers from various part of the world have started to work on casting prediction tools.

During the last thirty years, concrete has been industrially mutating from a soft granular medium to a proper non-Newtonian fluid. To benefit from the full potential of the modern fluid concretes such as Self-Compacting Concrete (SCC), prediction tools of the form filling taking into account the properties of the concrete, the shape and size of the structural element, the position of rebars, and the casting technique are needed. Although a lot of progress has been made in the field of fluid concretes, we must not forget that the most suitable concrete to cast a given element is still a concrete fluid enough to fill the formwork but not more. Additional and thus useless fluidity will always have a cost either in terms of super-plasticizer amount, loss of mechanical resistance or risk of segregation. Just as numerical simulations of concrete structures allow a civil engineer to target a minimum needed mechanical strength, casting prediction numerical tools could allow the same engineer to target a minimum workability that could ensure a proper filling of a given formwork.

Computational modeling of flow could therefore be used for simulation of *e.g.* total form filling and detailed flow behavior as particle migration and formation of granular arches between reinforcement ("blocking"). But computational modeling of flow could also be a potential tool for understanding the rheological behavior of concrete and a tool for mix proportioning. Progresses in the correlation between mix proportioning and rheological parameters would of course result but, moreover, the entire approach to mix proportioning could be improved.

Following the first international workshop organized in this field at CBI (Sweden) in September 2006, researchers from various research teams have realized that the numerical techniques they were using were almost as numerous as the researchers themselves. They decided to create a RILEM technical committee and had their first meeting in September 2007 in Ghent. During this first meeting I had the pleasure and honor to chair, these researchers decided to produce a state of the art report describing the present status regarding computational modeling of the flow of fresh concrete.

This report is divided into five chapters. The first chapter deals with the various physical phenomena involved in flows of fresh cementitious materials. The aim of the second chapter is to give an overview of the work carried out on simulation of flow of cement-based materials using computational fluid dynamics (CFD). This includes governing equations, constitutive equations, analytical and numerical solutions, and examples showing simulations of testing, mixing and castings. The third chapter focuses on the application of Discrete Element Method (DEM) in simulating the flow of fresh concrete. The fourth chapter is an introductory text about numerical errors both in CFD and DEM whereas the fifth and last chapter give some recent examples of numerical simulations developed by various authors in order to simulate the presence of grains or fibers in a non-Newtonian cement matrix.

I would like to finish this introduction by expressing my sincere gratitude to the contributors, without whom this RILEM report will never have been published

Claudia Bellmann
Ana Bras
Annika Gram
Bogdan Cazacliu
Dimitri Feys
Jesper Hattel
Jon Elvar Wallevik
Jon Spangenberg
Jörg-Henry Schwabe
Knut Krenzer
Ksenija Vasilic
Laetitia Martinie
Lars Thrane
Liberato Ferrara
Mette Geiker
Nathan Tregger
Nicolas Roquet
Nicolas Roussel
Nicos Martys
Paul Bakker
Samir Mokeddem
Sergiy Shyshko
Steffen Grünewald
Stephan Uebachs
Surendra Shah
Viktor Mechtcherine
Wolfgang Brameshuber

Contents

Chapter 1
Physical Phenomena Involved in Flows of Fresh Cementitious Materials

Nicolas Roussel[1] and Annika Gram[2]

[1] Université Paris Est., IFSTTAR, France
[2] CBI, Swedish Cement and Concrete Institute, Sweden

1.1 Introduction

The vast family of industrial cementitious materials presents such a variety of behaviors in the fresh state that describing them as a whole seems unattainable (*cf.* Fig 1.1). This is even more so for our objective here: studying the possibility to predict their response in practical processing conditions.

Our objective here is of course not to quantitatively predict rheology of cementitious materials but rather to understand situations where, depending on composition and processing, one or the other physical phenomenon (*e.g.* hydrodynamic interactions, contact forces, colloidal interactions…) will control the macroscopic rheological behavior. In this respect, we follow Coussot and Ancey [1], who proposed a conceptual diagram of predominant interactions in flowing concentrated suspensions under simple shear, as a function of shear rate and solid fraction. A similar work was carried out in the case of cement pastes and grouts in [2].

Although, in this chapter, we do not resolve or address detailed processing problems, we provide general guidelines and definitions to identify the numerous physical parameters which govern the macroscopic rheological (flow in the bulk) and tribological (flow at interfaces) behavior. In doing so, we build the basis on which numerical simulations of flow of fresh cementitious materials shall develop no matter the studied material, the specific scale of observation, the chosen numerical method or the type of industrial process of interest.

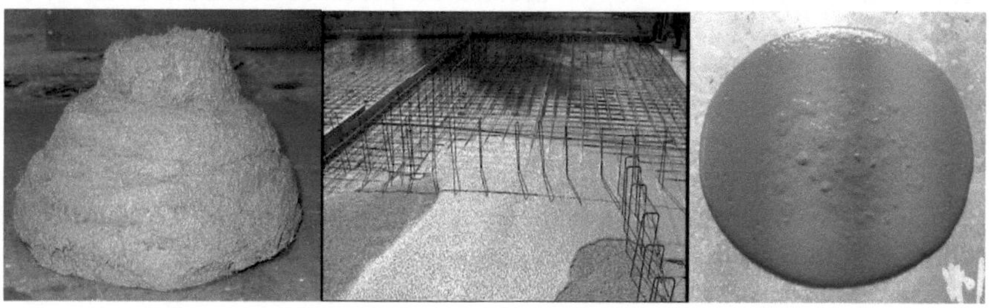

Fig. 1.1 Various behaviors in the fresh state. From left to right, Ultra High Performance Fiber Reinforced Concrete(UHPFRC) – Self Compacting Concrete (SCC) – Injection cement grout.

N. Roussel and A. Gram (eds.), *Simulation of Fresh Concrete Flow*,
RILEM State-of-the-Art Reports 15,
DOI: 10.1007/978-94-017-8884-7_1, © RILEM 2014

1.2 Is Concrete a Discrete or a Continuum Material?

As soon as one wishes to model the behavior of a given material in nature or industry, the question of the ratio between the scale of observation and the size of an elementary representative volume has to be dealt with [3].

Identifying concrete as a single fluid (or a continuum homogeneous medium) means that, in any two parts of the observed volume, we should find similar ensemble of components. If a physical quantity q (*for instance: velocity*) is studied, the minimum scale at which one can reasonably observe the system and effectively consider it as a single fluid thus corresponds to the point beyond which the average of q no longer varies when this scale further increases (see Fig. 1.2).

The volume corresponding to this scale of observation will be an elementary part of the material (as opposed to a component). However, it is worth noting that, when further increasing the scale of observation, macroscopic variations start to play a role (see Fig. 1.2). These variations are precisely what we want to model and predict. The appropriate scale of observation is thus situated between the range of rapid fluctuations of q due to matter discontinuity (*for instance: the variations in velocity between the aggregates and the cement paste*) and the range of macroscopic variations (*for instance the velocity field gradient in the formwork*). When these two ranges coincide, it is not possible to consider the system as a single fluid under the continuum assumption. In this particular situation, the flow is said to be in the "discrete" regime.

The **scale of observation** is thus of great importance to choose whether or not a single fluid approach is legitimate. The order of magnitude of a formwork smallest characteristic size is around 0.1 m while the order of magnitude of the size of the coarsest particles is around 0.01 m. This means that, if, as a first approximation, the presence of the steel bars is neglected, the flow in a typical formwork can be considered as the flow of a single fluid and a discrete modeling approach is thus not needed.

However, several reasons can lead researchers and engineers to the use of discrete methods or other alternative tools:

- If the objective is to understand the correlation between rheology and mix composition.
- If it seems necessary to take into account the presence of the steel bars or other obstacles as, in this case, the studied scale is of the same order as the size of the grains or if prediction of granular blocking is of interest.
- If prediction of segregation induced by the flow (*i.e.* dynamic segregation) is needed. Indeed, the single fluid techniques rely on the assumption that, of course, the medium is homogeneous but, moreover, that it stays homogeneous during flow. When segregation of the coarsest particles is the object of the study, it is thus necessary to introduce particles in the modeling.

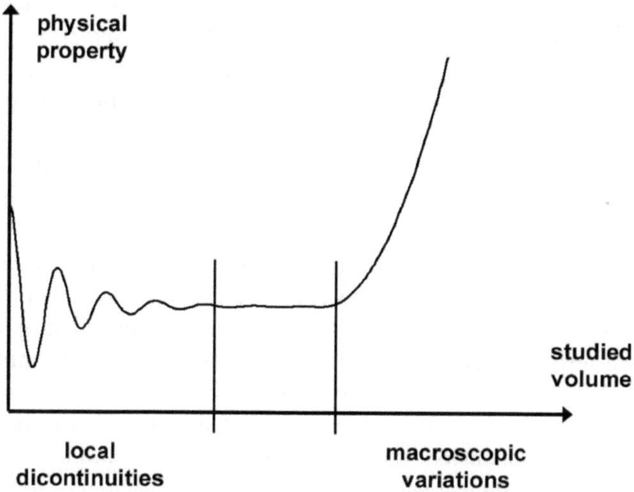

Fig. 1.2 Variation of a physical property averaged over a given volume of material as a function of the studied volume size

1.3 Macroscopic Rheological Behavior

Depending on their mix-proportioning and the considered range of shear rates, cementitious materials may display, in steady state flow, either **Newtonian** (*i.e.* constant apparent viscosity), **shear thinning** (*i.e.* decreasing apparent viscosity with shear rate) or **shear thickening** behavior (*i.e.* increasing apparent viscosity with shear rate).

Moreover, most cementitious materials display a **yield stress** that should be overcome to initiate flow from rest, and below which the flow stops during casting. The fact that fresh concrete displays a yield stress has a direct consequence on the shape of the mixture at stoppage during testing or casting of fresh concrete [4]. If we consider, for instance, the flow of a purely viscous fluid, the material would self level under the effect of gravity. Gravity would indeed induce a pressure gradient in the material if the upper surface of the material is not horizontal. This pressure gradient would generate a shear stress in the material that would create a shear rate and force the material to flow until the upper surface becomes horizontal and the pressure gradient at the origin of the flow has disappeared. The viscosity of the material would only play a role on the time needed to obtain a horizontal surface. As it will be reminded further in this report, yield stress is correlated with tests such as slump or slump flow [5].

Finally, another major aspect of the rheology of cementitious systems is its dependence on the flow history of the material. For example, the yield stress of a concrete left at rest or slowly sheared significantly increases with time. This leads to what is often described as "workability loss". Some of the structural changes responsible for the evolution of rheological properties are reversible: their effects are erased by mixing or by any type of strong shearing which brings the system back to a previous structural state. These reversible changes, often dominant on short observation times, are generally described as **thixotropy** [6-8]. Other changes, in particular some consequences of the hydration phenomenon, are irreversible [9, 10] and thus contribute to the long term evolution of material properties (towards the solid state) even if they take place on longer time scales than thixotropy.

It can be noted that Tattersall in early works [11] did not register a full recovery of the initial structure after a strong shear (*i.e.* a reversible evolution of the behavior). He was therefore very specific on calling this a non-thixotropic process, and therefore the term "structural breakdown" was preferred over "thixotropic behavior" [12-14]. This structural breakdown was attributed to the process of breaking certain links due to early hydration products between the cement particles. During stirring of cement suspension, both thixotropic and structural breakdown behavior may be occurring simultaneously [15].

1.4 Multi-scale Approach

Concrete components range from micrometric cement particles to centimetric aggregate particles. In order to overcome this difficulty in the study of the relation between mix design and rheological properties, multi-scale approaches seem very promising [16-18]. The simplest **multi-scales approach** only involves two scales and it is therefore often considered that concrete is a suspension of aggregate particles (sand and gravel) in a suspending fluid (cement paste). It can then be expected, as a first approximation that, as for many suspensions in nature or industry, the rheological properties (*i.e.* yield stress or viscosity) of the suspension (*i.e.* the concrete) is proportional to the yield stress of its suspending fluid (*i.e.* the constitutive cement paste).

This statement can be extrapolated from general relations similar to Krieger-Dougherty relation [19-21] for apparent viscosity, which relates the rheological properties of the suspending fluid and the volume fraction ϕ of the particles to the rheological properties of the mixture. It must however not be forgotten that this type of phenomenological relation has been historically established for purely hydrodynamic interactions between the suspending fluid and the particles and thus should only apply for moderate inclusions volume fraction.

However, these approaches could be subject to debate as there does not exist any reason to consider that cement paste is the suspending fluid more than water or mortar. Let us describe the consequences of the three types of approaches which can be found in literature on the modeling of the mixture behavior.

- Water is the fluid

Water may be considered as the suspending fluid and all particles in the mixture are considered as inclusions and participate to the total solid content of the mixture [22-25]. This is the most rigorous approach from a theoretical point of view. However, difficulties come from the fact that all types of inter-particles interactions (from frictional interactions between the largest grains to the highly complex colloidal interactions occurring at the cement paste scale) are to be taken into account in order to describe the system. The knowledge of the total solid content is not sufficient to know which type(s) of inter-particles interactions will dominate. As each type of interaction is more or less associated to a particle size [2, 17], any quantitative predictions can only come from a distinction between the various species in the mixture. Moreover, within the frame of numerical simulations of flow of concrete, this would mean that the number of particles to be simulated would be tremendously high and would prevent any industrial application of the numerical technique to casting process prediction.

- Cement paste is the fluid

The natural answer to the above difficulties consists in considering, as described above, that concrete is a suspension of aggregate particles (sand and gravel) in a suspending fluid (cement paste). Two granular species are thus considered: more or less colloidal particles and non-colloidal particles. One has then to assume that the ratio between the aggregate particles diameter and the cement particles diameter is far higher than 1. If the diameters of the largest particles are considered (*i.e.* 100 μm for the cement particles and around 10 mm for the coarse particles), cement paste can thus be considered as a continuum medium and therefore as the suspending fluid at the scale of the sand and gravel particles. However, sands and gravels also contain small amount of fine particles, the diameter of which is of the order of the largest cement particles. Choosing this approach thus needs to neglect this overlap in species diameter. Moreover, it has been shown recently that aggregate particles (probably the finest with the largest specific surface) may adsorb a non negligible part of the High Range Water Reducing Admixtures (HRWRA) [26]. This means that the independently measured cement paste yield stress or viscosity is systematically lower than its yield stress or viscosity once inside concrete and that the calculated relative yield stress or viscosity will be a strong approximation of the real relative yield stress or viscosity of the material studied.

- Mortar is the fluid

The last type of approach that can be found in literature consists in considering that the inclusions are the aggregate particles larger than a given separation size. The question that arises is then “which size?”. Several values can be found in the literature between 0.125 mm and 2 mm [18, 27-29]. This approach is of course more rigorous from a theoretical point of view as all the potentially colloidal particles can be studied together but it is also more complicated from a practical point of view as it requires to isolate the finest sand particles and add them to the cement paste to be studied.

1.5 Particle Interactions

1.5.1 Review of Interactions

Despite the variety of parameters and the complexity of physical phenomena governing macroscopic behavior, it is possible to identify a few dominating phenomena governing the material response.

As stated above, fresh cementitious materials can be viewed as suspensions of particles of many different sizes (from several nm to a few cm) in a continuous fluid phase. This broad poly-dispersity implies that various interactions interplay. We can at least identify four main types of interactions:

- Brownian forces
- Surface forces (or colloidal interactions)
- Hydrodynamic forces
- Various contact forces between particles.

Depending on the size of the particles, on their volume fraction in the mixture and on external forces (e.g. the magnitude of either the applied stress or strain rate), one or several of these interactions dominate [2, 17].

1.5.2 Brownian Forces and Colloidal Interactions at the Cement Paste Scale

As it is very rare in literature and in practice to try to simulate or model the flow at this scale of observation, we will not dwell here on details. We shall however remind the following global scheme:

Because of Brownian motion, cement particles diffuse through the liquid and may get close enough for van der Waal attractive forces to strongly increase and dominate electrostatic repulsion. As Brownian motion can be neglected compared to short range van der Waal attractive forces, it is not able to separate the particles and they agglomerate. It is crucial to keep in mind that network formed by agglomerated particles is at the **origin of the yield stress** of cementitious materials. This network may be very strong and may prevent the mixture from flowing properly. In order to prevent this situation from occurring, water reducing admixtures may be used. They can be seen as molecules which get adsorbed at the surface of cement grains and, through steric hindrance or electrostatic effects, increase the average inter-particle distance and therefore decrease the magnitude of the attractive van der Waal forces and the strength of the aggregated particles network (*cf.* Fig. 1.3).

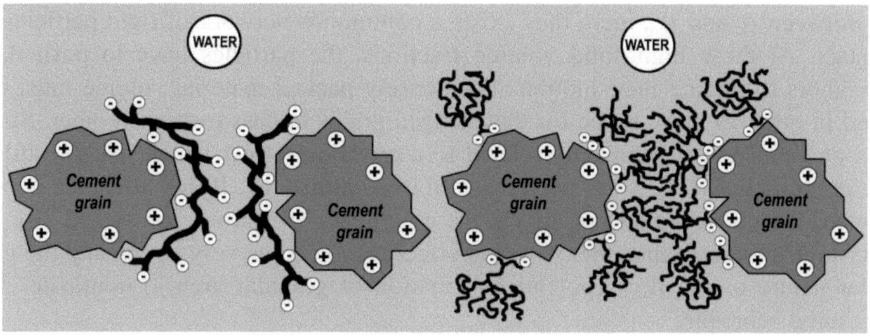

Fig. 1.3 Effects of HRWRA on flocculated cement grains. By increasing the distance between cement grains at contact points either through the introduction of electrostatic repulsive forces (left) or steric forces (right), HRWRA reduce the intensity of the attractive Van der Waals forces in the network of flocculated particles. They therefore allow for a decrease in the yield stress of the suspension.

1.5.3 Direct Contact Network between Particles

An analogy is often made between the flow of liquids and that of granular media, even though the physical properties of the two are quite different. Concrete by nature is dominated by its fluid-like behavior or by its granular media like behavior according to its mix design, casting geometry and casting process. In the case of SCC for example, the amount of coarse particles in the mixture is low and this modern concrete behaves as a fluid suspension whereas, in the case of ordinary concrete with greater amount of coarse particles, the behavior is dominated by the granular nature of the material [30]. It will be shown further in this report how Distinct Element Methods (DEM), which were initially developed to model the flow of dry granular media, can be used to model flow of concrete.

It is known that stable packing of rigid particles is observed only above a critical packing fraction ϕ_c, which can be associated to a percolation process of particles in contact (*cf.* Fig. 1.4). Experimental and numerical results indicate that, for uniform spheres, ϕ_c (the so-called **Random Loose Packing** (RLP) fraction) should be situated around 0.5 [31]. There exists also a **dense packing fraction** ϕ_m (the close packing fraction), which is a geometric property of packing of non-interacting rigid particles (*cf.* Fig. 1.4). Both ϕ_c and ϕ_m are directly related to the poly-dispersity and shapes of the constitutive particles. For example, for identical spheres, the random close packing ϕ_m is of the order of 0.64.

Note that both ϕ_m and ϕ_c should increase in the case of poly-disperse particles. In fact, experimental observations [26, 27, 30] even suggest that the ratio and ϕ_c/ϕ_m (equal to 0.85 for mono-disperse spheres) is roughly constant for many different poly-dispersity and particle shapes. This can be understood as the average distance between particles given scales with $(\phi/\phi_m)^{-1/3}$ [22] and hence packing properties primarily depends on (ϕ/ϕ_m).

Between ϕ_c and ϕ_m, there thus exists a continuous network of rigid particles in contact. At these high solid volume fractions, the particles have to push their neighbors to initiate their motion. The densely packed material volume must expand in order to make space for the encaged grains to pass over each other. Since this phenomenon is intimately related to a crowding effect which tends to dilate the granular network, it has been referred to as **dilatancy**. If this dilatancy is prevented, the crowding shows up via the emergence of normal forces [32-33]. At very high solid contents, the crowding effect may become very strong and the granular nature of the flow (potential formation of granular arches) dominates the measured behavior.

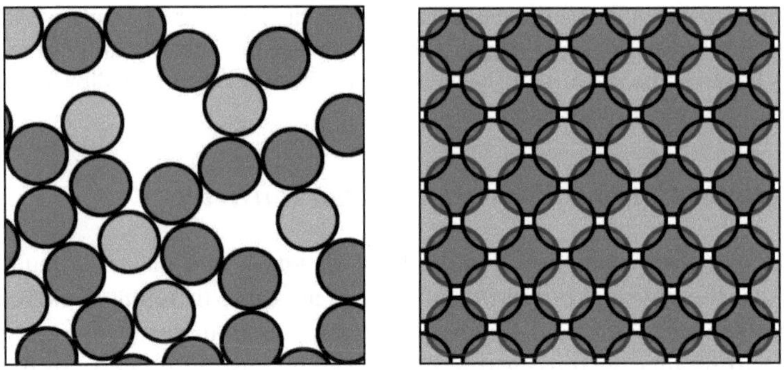

Fig. 1.4 Packing of rigid particles. (left) random loose packing fraction for which a contact network appears in the granular materials (of order 0.55 for uniform spheres). (right) dense packing fraction (of order 0.64 for uniform spheres).

1.5.4 Hydrodynamic Interactions and Viscosity

It is common for many suspensions to present a macroscopic viscous behavior over some intermediate range of $\dot{\gamma}$. Concrete for instance displays above the yield stress a viscosity called plastic viscosity in reference to the Bingham model, that finds its origin in the viscous dissipation in the interstitial fluid. Even in the case of the most simple mix designs, viscosity is a complex function of the volume fraction. For low values of ϕ, say, simple analytical relations such as Einstein relation can be applied:

$$\mu = \mu_0(1+2.5\phi) \tag{1}$$

where μ is the apparent viscosity of the suspension and μ_0 is the viscosity of the suspending fluid. The above relation holds in general for volume fractions lower than 5 %, which, in the case of cement pastes, corresponds to a water to cement mass ratio W/C higher than 6, a situation which is never encountered in practice.

For larger values of ϕ, experimental observations show that the apparent viscosity deviates significantly from the Einstein relation and eventually diverges when the solid volume fraction tends towards the dense packing fraction ϕ_m. Various empirical expressions have been proposed, the most famous one being the Krieger Dougherty equation [19]. The general form of all these equations is the following:

$$\mu = \mu_0 \left(1 - \frac{\phi}{\phi_m}\right)^{-q} \tag{2}$$

The specific microscopic processes responsible for this behavior are still a matter of debate. Clearly, as ϕ increases, at intermediate volume fractions, energy dissipation progressively concentrates in strongly sheared fluid layers between neighboring particles. Moreover, above ϕ_c (see previous section), actual particle/particle contacts exist and contribute to the macroscopic stress. The value of the viscosity near ϕ_m hence results from the interplay between hydrodynamics inside the interstitial fluid and occasional contacts between particles. For more on this topic, see [34].

The specific values of q and ϕ_m are also matter of debate. In its most famous form, the Krieger-Dougherty equation, Eq. (2) is written with $q=2.5\phi_m$ for spheres so that it matches Eq. (1) at low volume fractions while capturing the divergence near ϕ_m. Recent studies suggest instead that q takes on the simple value $q=2$ for spheres [34]. The actual value of q may also depend on particle shape, some studies suggesting $q>2$ for non-spherical particles. This question however requires further studies using modern rheological tools.

It may be worth here to bring into perspectives a theory called "Excess paste theory" [35], which is strongly spread within the concrete community. The concept of a layer of cement paste surrounding all sand and gravel particles with varying thickness according to the granular content is often used to explain the influence of coarse aggregate content on workability. This layer is said to lubricate the grains relative movement and increase the flowability of the mixture. This concept however is far from what can be found in more fundamental research areas, in which solid or inclusions volume fraction ϕ is the only parameter used to describe the system jamming state and thus its ability to deform [1]. The concept of thickness layer has proved its efficiency in explaining many observed phenomena [36, 37]. It can however be noted that paste excess thickness is in fact another way to express the average distance b between granular inclusions of diameter d which writes:

$$b = d\left(1 - (\phi/\phi_m)^{-1/3}\right) \tag{3}$$

It can be noted that Eq. (3) is only a very rough approximation of an average inter-particles distance. For a more realistic calculation of a distribution of inter-particles distance, one should read [38].

The fact that a correlation exists between the excess paste layer thickness calculated from Eq. (3) and the rheological properties of concrete can therefore be explained by the fact that there exists a relation between rheological properties and jamming of the system. This jamming is correlated to the inclusions volume fraction. Moreover, there exists a direct relation between volume fraction and average distance between aggregate inclusions. Finally, this average distance is equal to two times the cement paste excess layer (see Fig. 1.5).

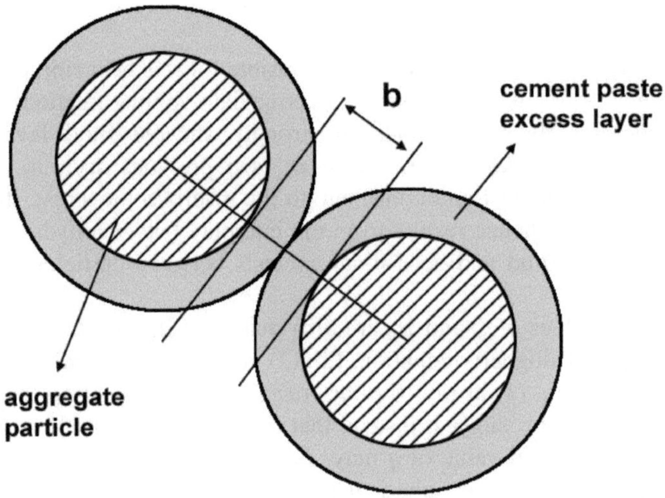

Fig. 1.5 Cement paste excess layer and average distance between particles

1.5.5 Relative Contributions of Yield Stress and Viscosity and Bingham Number

It may be worth noting here that, as viscous contributions depend on shear rate whereas colloidal contributions do not, the relative contribution of colloidal interactions and viscous dissipation depends obviously on the material but also on shear rate, which itself depends on casting process. Yield stress is therefore the most important parameter when the way flow stops or starts is of interest. At higher shear rates, however, plastic viscosity may become dominant.

One common way to quantify this relative contribution is the computation of what is called the Bingham number. This dimensionless number is the ratio of the yield stress contribution to the viscous contribution. It can be expressed as:

$$B_n = \frac{\tau_0}{\mu_p \dot{\gamma}} \cong \frac{\tau_0 D}{\mu_p V} \tag{4}$$

Where τ_0 is the yield stress of the concrete, μ_p is the plastic viscosity, D is the typical dimension of the flow and V is the flowing velocity. If we consider an ordinary rheology concrete with a yield stress of the order of 1000 Pa and a plastic viscosity of 100 Pa.s being cast in a 20 cm wide formwork at an average velocity of 0.05 m/s, the Bingham number is approximately 40, showing that yield stress dominates flow. On the other hand, if we consider a Self Compacting Concrete with a yield stress of 50 Pa and a plastic viscosity of 150 Pa being pumped at a flowing velocity of 1 m/s in a pipe of diameter 12 cm, the Bingham number is of the order of 0.04 showing that plastic viscosity becomes the most important parameter.

1.5.6 Kinetic Energy and Reynolds Number

Most flows in industry are considered as being laminar. This means that the kinetic energy of the system can be neglected compared to all other sources of dissipation. It may be useful to keep in mind that it is not always the case. Whereas the Bingham number described above is the ratio of the yield stress contribution to the viscous contribution, the Reynolds number is the ratio of the kinetic energy to the viscous energy. It writes:

$$R_e = \frac{\rho V D}{\mu} \tag{5}$$

Where ρ is the density of the material, μ is viscosity, D is the typical dimension of the flow and V is the flowing velocity In the case of a yield stress fluid such as concrete, the apparent viscosity is often considered as viscosity in Eq. (5):

$$\mu_{app} = \frac{\tau_0}{\dot{\gamma}} + \mu_p \tag{6}$$

Where τ_0 is the yield stress of the concrete and μ_p is the plastic viscosity. The Reynolds number becomes:

$$R_e = \frac{\rho V^2 D}{D\tau_0 + \mu_p V} \tag{7}$$

Let us reconsider the two above examples. First, we consider the ordinary rheology concrete with a yield stress of the order of 1000 Pa and a plastic viscosity of 100 Pa.s being cast in a 20 cm wide formwork at an average velocity of 0.05 m/s. The Reynolds number is approximately 0.006, showing that the flow is indeed largely laminar. On the other hand, if we consider the Self Compacting Concrete with a yield stress of 50 Pa and a plastic viscosity of 150 Pa being pumped at a flowing velocity of 1 m/s in a pipe of diameter 12 cm, the Reynolds number is of the order of 2 showing that turbulent effects could start to play a role.

1.6 Stability and Static Segregation

The densities of the numerous components entering concrete mix proportions vary between 1000 kg/m^3 (water) and 3200 kg/m^3 (cement). Even materials lighter than water may be used in the case of lightweight concrete. With such a mixture, gravity quickly becomes the enemy of homogeneity.

In the field of cementitious materials, heterogeneities induced by gravity are divided in two categories depending on the phase that is migrating: bleeding and segregation. Both are induced by the density difference between the components but the bleeding phenomenon is concerned with the water migration whereas segregation is concerned with the movement of the coarsest particles.

As we mainly focus in this report on the flow of concrete, we will therefore deal here with the segregation of the coarsest particles.

As described in the section dealing with multi-scale approaches (section 1.3), segregation in mortars may then be predicted from paste rheology and the volume fraction, shape and size distribution of the sand particles whereas segregation in concrete may then be predicted from mortar rheology and the volume fraction, shape and size distribution of gravel. Of course, segregation of both sand and gravel may also be predicted from paste rheology and volume fraction, shape and size distribution of all coarse particles. The physical concept of inclusions in a suspending fluid is adapted to the description of each of the constitutive phase of the concrete and was already used in several papers [39, 40] to predict stability of SCC.

Attempts to correlate rheology of fresh concrete to its stability can be found in the literature [41-43]. Figures linking either empirical test results or even quantitative measurement of segregation to the slump or the slump flow of a given concrete have often been plotted. Although a correlation may, in certain cases, be obtained, no information about the underlying physics will ever be gathered this way. Segregation is a multi-phase separation phenomenon (the minimum number of phases is two: a suspending fluid and some solid inclusions). As such, the only relevant approach is a multi-phase one. The rheological behavior of concrete has no role to play; only the rheological behavior of the suspending fluid does matter. When correlations between concrete behavior and its stability are obtained, it is only because concrete behavior is strongly linked to the behavior of its suspending fluid.

Segregation (or stability property) is often associated to static segregation. The material is not flowing and the particles "fall" towards the bottom of a given sample or of a formwork as their density is higher than the density of suspending fluid. The involved physical phenomena are of course the same if the material is flowing. However, other reasons than gravity such as strain rates gradient may induce or enhance segregation. Segregation, when the material is flowing is called "dynamic segregation".

In the case of a cement paste at rest behaving as a fluid with a yield stress τ_0, it was shown in [40] that, if interactions between particles or any lattice effects [44] are neglected, particles smaller than a critical diameter d_c are stable while larger ones may segregate. d_c may be computed as :

$$d_c = \frac{18\tau_0}{\Delta\rho g} \tag{8}$$

where $\Delta\rho$ is the density difference between the suspended inclusions and the cement paste.

Finally, it should be kept in mind that, according to the above criterion, static stability of a given mixture depends on the yield stress of the suspending fluid but not on its viscosity. The plastic viscosity has of course an influence on the separation velocity if the particle is not stable but does not affect whether or not the particle is stable. One could object that, if the plastic viscosity is high enough, then, even if the particle is not stable, the separation velocity will be so low that the displacement of the particles before setting will be negligible. To have an idea of the order of magnitude of the plastic viscosity that would be needed to prevent a sand grain (diameter 4 mm, density 2700 kg/m^3) from settling in a cement paste (density 1900 kg/m^3), let us do the following calculation: we assume that segregation is negligible if the sand particles displacement before setting is less than 1 mm and that setting occurs after 4 hours. The settling velocity of the 4mm sand particle should then be lower than 7.10^{-8} m/s. The viscosity needed to reach this velocity may then be compute from Stokes Law and is higher than 9000 Pa.s, which is at least 9000 times higher than a standard cement paste viscosity. Of course, one could try to find equilibrium between yield stress and viscosity, to get a quasi-stable mixture but, most of the time, the chosen yield stress will have to be very close to the critical yield stress that fully prevents a sphere from moving under its own weight. In order to design stable fluid concrete with lower values of the constitutive cement paste yield stress, it is more interesting, as it will be discussed further, to rely on structural build up of the cement paste.

1.7 Dynamic Segregation and Granular Blocking

Although the fluidity of a given concrete could allow for the filling of formworks with complex shapes, it should not be forgotten that concrete contains coarse particles that could get jammed in the most reinforced zones during the casting process (see Fig. 1.6). When concrete flows through an obstacle such as steel bars, several phenomena occur between which a distinction should be made.

As already described in the second section of this chapter, if the shear stress generated by gravity, which is a complex function of the flow geometry, becomes lower than the yield stress of the concrete, flow stops. This effect has been quantified in the case of the L-Box test with and without steel bars [45] and it was demonstrated that the concrete level thickness variation between the case with bars and without bars is of the order of $3\tau_0/\rho g$ where τ_0 and ρ are respectively the yield stress and the density of the tested SCC. For traditional SCC, the yield stress of which is of the order of 100 Pa, this variation is therefore of the order of 1 cm. This value was validated by testing stable limestone filler suspensions, which did display a yield stress of the same order as SCC, but the constitutive particles of which were too small to create a granular blocking in the vicinity of the obstacle. This also explains why there exists, even for stable concretes which do not display any granular blocking, a systematic difference between slump flow and J-Ring test [46, 47]. It is interesting to note here that, in [47], all the measurements and conclusions may be explained by the yield stress variation between the various tested concretes and that there is no granular blocking at all.

Fig. 1.6 Granular blocking during a L-Box test on SCC (from [48])

Moreover, as described in the previous section, the coarsest particles in concrete are subjected to gravity and are immersed in a fluid with a lower density, and of viscosity possibly too low to prevent them from settling or segregating within the flow duration. If concrete is at rest, it has been shown that it is the yield stress of the cement paste that may prevent these coarse particles from settling. When concrete is flowing, the drag force exerted by the suspending fluid (mortar or cement paste depending on the multi-scales frame chosen) on each particle has to be high enough to "carry" the particles. If the studied concrete is not stable, then the presence of the obstacle could increase segregation effects. Indeed, it is a known feature of suspensions that particles migrate from high shear rates zones to lower shear rates zones [34]. The flow perturbations induced by steel bars locally increase these shear rate gradients and can thus increase shear induced segregation. It can be noted that, according to the knowledge of the authors, this phenomenon has still not been modeled and properly measured from a practical point of view in the case of concrete. Although the above phenomena do not lead directly and systematically to granular blocking, they can strongly affect the coarsest particles configuration and volume fraction at the vicinity of the obstacle.

Finally, if the characteristic size of the obstacles (*e.g.* the gap between the bars) is not far from the size of the coarsest particles, proper granular blocking may occur and granular arches may appear, stable enough to resist the flow. At the origin of the formation of these granular arches is granular clogging, jamming or blocking, namely the fact that the suspended particles at some time jam somewhere in the obstacles formed by the steel reinforcements. It can be noted that granular blocking may occur for particles with a diameter smaller than the gap between obstacles and that it is thus essentially a collective effect. It has however to be kept in mind that segregation induced by flow or gravity and described above may lead to an increase in the local volume fraction of coarse aggregates which could itself locally increase the risk of granular blocking.

The following striking points about granular blocking of a suspension can be gathered from [48]:

- The granular blocking phenomenon has a probabilistic nature: for a given experiment, the measured residues vary significantly according to the specific, initial distribution of particles in the fluid, which cannot be controlled; this emphasizes that, at a local scale, the granular blocking is related to the probability of presence of particles. As a consequence, it should not be possible to completely suppress the risk of granular blocking in a given concrete casting process but only to reduce it below an acceptable level.
- From a probabilistic point of view, a granular blocking event requires that the particles be sufficiently close to each other, and thus is more probable for large particles volume fraction. This result in particular means that it should be possible to improve casting process by adjusting the coarsest inclusions volume fraction and not only the diameter of the coarsest particles. This means additionally that, as SCC contains less coarse inclusions than Ordinary Rheology Concrete (ORC), its probability of granular blocking (or passing ability) should be lower (higher) than ORC.
- The number of granular blocking events increases with the number of attempts (drawings, in probabilistic terms), which implies that, even if the probability of granular blocking is low, the residue will increase with the crossing volume for a given particle volume fraction. From a practical point of view, this means that the volume of fresh concrete that has to cross the obstacle should play a strong role on the probability of blocking.
- The rheological behavior of the suspending fluid does not seem to play any role in the blocking phenomenon. As a consequence, this means that granular blocking during concrete casting should only depend on the number and size of the coarsest particles crossing the obstacle, their volume fraction and the free spacing between the bars but not on the rheological behavior of the constitutive cement paste. This, of course, stays true only if the cement paste is viscous enough to prevent any dynamic segregation of the concrete which could locally increase the volume fraction of the particles [49].

Finally, it may be useful to insist here on the fact that the rheology of the cement paste should not affect the granular blocking phenomenon. It is indeed natural to imagine that a low viscosity cement paste and thus a very fluid SCC would be more prone to have its coarsest particles blocked in highly reinforced zones. However, as stated above, the fact that the material is too fluid to carry its own particles during flow is not directly at the origin of the granular blocking. The consequence of this high fluidity is that the material is not stable and that the local coarse particles volume fraction may increase above the volume fraction deduced from mix proportions. This increased volume fraction due to segregation may thus be sufficient to create granular blocking although the volume fraction deduced from mix proportions was not.

1.8 Fiber Orientation and Induced Anisotropy

Contrary to spherical rigid inclusions, in the case of rigid fibers, flow can induce a preferred orientation of the fibers (*cf.* Fig. 1.7) which modifies the material fresh properties and, after setting, strongly influences the mechanical properties of the resulting fiber-reinforced composite [50-54].

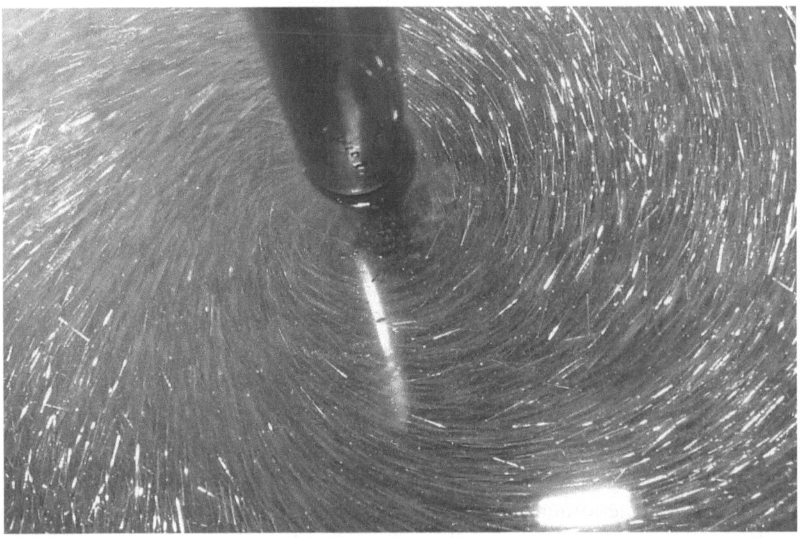

Fig. 1.7 Fiber orientation in oil

Most studies on fibers suspensions owe their foundations to the solution by Jeffrey [55] for the motion of a rigid ellipsoid suspended in a Newtonian fluid experiencing simple shear flow. Jeffrey shows that a lone rigid, force and torque free ellipsoid will translate with a velocity equal to that of the equivalent undisturbed fluid at the ellipsoid's centre. Furthermore, the ellipsoid's axis of revolution will orient itself in a repeating and characteristic motion commonly referred as Jeffrey's orbit. The Jeffrey's orbit is a periodic and unchanging trajectory around the flow vorticity axis characterized by long periods aligned in the shear direction, broken by brief reversals, whereby the ellipsoid's main axis essentially flips over in a tumbling fashion until the opposite end is again aligned in the shear direction.

In the absence of a disturbance, such as an interaction with a neighboring fiber, the ellipsoid will remain on this orbit and the period T between two brief reversals may be computed as:

$$T = \frac{2\pi}{\dot{\gamma}}\left(r + \frac{1}{r}\right) \tag{9}$$

where $\dot{\gamma}$ is the shear rate and r is the aspect ratio of the fiber (*i.e.* ratio between the length of the fiber and its diameter). The higher the fiber aspect ratio is, the stronger anisotropy is induced by a shear flow.

In the dilute regime, the orientation does not depend on the fiber volume fraction according to Jeffrey's theory and high aspect ratio fibers tends to align better with the flow. It should be noted that, in dilute regime, if sufficient time is given to the fiber, full anisotropy will appear. This needed time is of the same order as the period T. After this transition time, the orientation vector stays on the Jeffrey's orbits.

However, in the case of cementitious materials, the amount of fibers added to the suspension is most of the time sufficient to create interactions between fibers such as direct contacts or hydrodynamic interactions. No equations such as Jeffery equation exist to describe the effect of these additional interactions on the orientation process. It can however be imagined that, when the length of the fiber is of the same order as the maximum aggregate size, fiber orientation will be dictated be the random contact interaction between grains and fibers and will therefore become itself random (*i.e.* isotropic).

1.9 Thixotropy and Transient Behavior

As long as steady state flow is reached, behavior of most cementitious materials may be described using yield stress models or at least properties which do not depend on time. However, between two successive steady states, there exists a transient regime, during which most steady state models are not sufficient to describe the observed behavior.

Let us consider a typical torque measurement obtained from a concrete rheometer during an instantaneous rotating speed decrease (Fig. 1.8(a)) or increase (Fig. 1.8(b)). The bold line shows what should be expected from a simple yield stress fluid whereas the thin line shows the real measurement obtained in practice. The difference is due to the thixotropic behavior of the tested concrete that creates a delay in the material answer. It has been shown recently in [56] that this delay in the case of cement pastes can be correlated to the applied shear rate and to the recent flow history of the material. If the material is at rest before a low shear rate is applied to the sample, a typical Vane test answer is obtained as shown in Fig. 1.8(c). After a linear increase due to the elastic part of the behavior, a static (or apparent) yield stress τ_0 can be measured that increases with the resting time before the test whereas the dynamic (or intrinsic) yield stress τ_{00} corresponding to steady state does not depend on the material flow history.

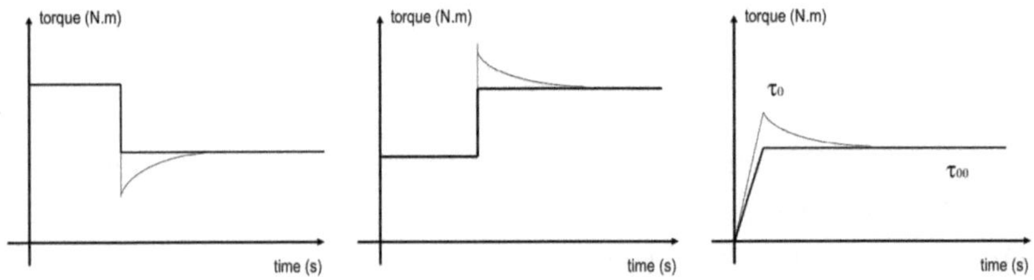

Fig. 1.8 Examples of transient flow behaviors. (left) rotating speed decrease;(center) rotating speed increase; (right) rotating speed increase after a resting period.

However, in the case of cementitious materials, things are not so simple as the hydration process starts as soon as cement and water are mixed together. The apparent viscosity of the material is permanently evolving as described in [7-15]. Recently, Jarny and co-workers [57] have however shown using MRI velocimetry that, over short timescales, weak aggregation processes dominate, which lead to rapid thixotropic (reversible) effects, while over larger timescales, hydration induces shear resistant aggregation, which lead to irreversible evolutions of the behavior of the fluid. These two effects might in fact act at any time but they appear to have very different characteristic times. As a consequence, it is reasonable to consider that there exists an intermediate period, say around a couple thousands seconds, for which irreversible effects have not yet become significant. This means that it seems possible to model thixotropy and only thixotropy on short periods of time (not more than 30 minutes as an order of magnitude) during which the irreversible evolutions of the concrete can be neglected.

A non-exhaustive list of the applications of such a model that are *a priori* concerned by thixotropy could be the following:

- Self Compacting Concrete (SCC) formwork pressure: during placing, the fresh SCC behaves as a fluid but, if cast slowly enough or if at rest, it flocculates and builds up an internal structure and has the ability to withstand the load from concrete cast above it without increasing the lateral stress against the formwork.
- Multi-layer or distinct layers casting: during placing, a layer of SCC has a short time to rest and flocculate before a second layer of concrete is cast above it. If it flocculates too much and its apparent yield stress increases above a critical value, then the two layers do not mix at all and, as vibrating is prohibited in the

case of SCC, this creates a weak interface in the final structure. Loss of strength of more than 40 % has been reported [58].
- Stability of SCC: during placing, the cement paste is de-flocculated because of the mixing and of the casting itself. This allows an easy placement of the material. However, as soon as casting is over and before setting, gravity may induce sedimentation of the coarsest particles as described above. A thixotropic cement paste will flocculate once at rest. Its apparent yield stress will increase and will be sufficient to prevent the particles from settling.

1.10 Behavior at the Walls

1.10.1 Slip Velocity and Slip Layer

Depending on the scale of observation [3], either a slip layer or a slip velocity may be observed. At large scales of observation (macroscopic level), an apparent slip velocity may be measured whereas, at smaller scales of observation, a slip layer appears, in which the velocity evolves from zero at the wall to the apparent slip velocity at the boundary of the slip layer (see Fig. 1.9).

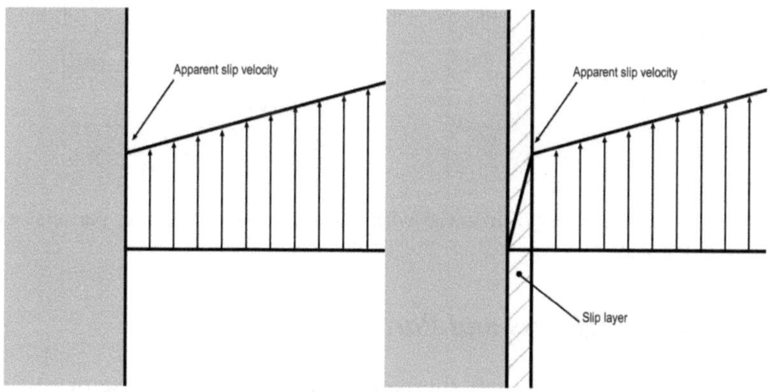

Fig. 1.9 Apparent slip velocity and slip layer. (left) macroscopic scale (right) microscopic scale.

1.10.2 Wall Effect

In the case of heterogeneous materials such as concrete, all slip is considered as resulting from a decrease in the concentration of the constitutive elements at the vicinity of the wall. The natural depletion of particles, also known as wall effect, prevents any particles with a diameter d from being located at a distance smaller than $d/2$ from the wall (see Fig. 1.10).

The mean concentration of inclusions being lower at the vicinity of the wall, the local viscosity of this less concentrated material is also lower. As a consequence, at a given shear stress, the shear rate becomes higher, creating locally a discontinuity and therefore a slip layer. At a macroscopic scale, this may be measured as a slip velocity.

When flow is steady and in the case of strongly poly-disperse materials such as concrete, this phenomenon may be enhanced by particle migration under shear. This phenomenon however needs time to develop and, although no quantification of this time has ever been done, it seems limited in practice to steady fast flows such as pumping or test in cylindrical rheometers.

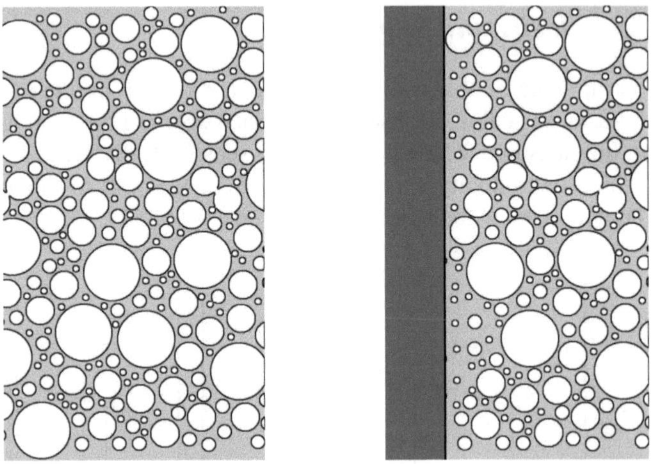

Fig. 1.10 Wall effect. A layer of material with a lower content of coarse particles exists at the vicinity of the wall.

1.10.3 Wall Roughness and Particles Size

It is generally assumed that wall slip can be prevented by the use of sufficiently rough surfaces. Indeed, as shown in Fig. 1.11, rough surfaces allow for an homogeneous (at least initially) distribution of the particles at the wall. The roughness needed to prevent any slippage is however subject to debate. It should be in the case of poly-disperse materials of the order of the size of the coarsest particles. In the case of concrete, it is the case only in concrete rheometers where roughness of the centimeter scale is used most of the time. During casting or testing, the surfaces are most of the time visually smooth. When measured, the roughness of steel formwork is of the order of 10 μm [59]. This means that formwork surfaces may be considered as rough at the cement particles scale whereas they may be considered as smooth at the sand and gravel scale.

As a consequence, it may be often considered that, as long as the flow is not steady and that no coarse particle migration occur and create a slip layer, it seems possible to consider that cementitious materials more or less stick to interfaces. This of course may prove wrong if these interfaces are lubricated by demoulding agent although this research area is still much opened.

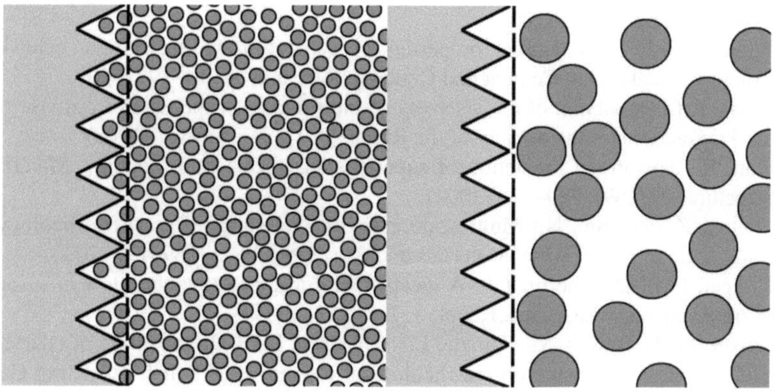

Fig. 1.11 Effect of wall roughness. (Left): no slip condition. (Right): slippage.

References

[1] Coussot, P., Ancey, C.: Rheophysical classification of concentrated suspensions and granular pastes. Phys. Rev. E 59(4) (1999)

[2] Roussel, N., Lemaître, A., Flatt, R.J., Coussot, P.: Steady state flow of cement suspensions: A micromechanical state of the art. Cement and Concrete Research 40, 77–84 (2010)

[3] Coussot, P.: Rheometry of pastes, suspensions and granular materials. Published by Jon Wiley and sons, Inc., Hoboken (2005)

[4] Roussel, N.: Rheology of fresh concrete: from measurements to predictions of casting processes. Materials and Structures 40(10), 1001–1012 (2007)

[5] Roussel, N., Coussot, P.: "Fifty-cent rheometer" for yield stress measurements: from slump to spreading flow. J. Rheol. 49(3), 705–718 (2005)

[6] Coussot, P.: Rheometry of pastes, suspensions, and granular materials. John Wiley & sons, New Jersey (2005)

[7] Tatersall, G.H.: Structural Breakdown of cement pastes at constant shear rate. Nature 175, 166 (1955)

[8] Roussel, N.: Steady and transient flow behaviour of fresh cement pastes. Cement and Concrete Research 35, 1656–1664 (2005)

[9] Banfill, P.F.G., Saunders, D.C.: On the viscosimetric examination of cement pastes. Cement and Concrete Research 11, 363–370 (1981)

[10] Otsubo, Y., Miyai, S., Umeya, K.: Time-dependant flow of cement pastes. Cement and Concrete Research 10, 631–638 (1980)

[11] Tattersall, G.H., Banfill, P.F.G.: The Rheology of Fresh Concrete, ch. 2 and 16. Pitman Books Limited, Great Britain (1983)

[12] Banfill, P.F.G.: The Rheology of Fresh Mortar. Mag. Concr. Res. 43(154), 13–21 (1991)

[13] Banfill, P.F.G., Saunders, D.C.: On the Viscometric Examination of Cement Pastes. Cem. Concr. Res. 11(3), 363–370 (1981)

[14] Banfill, P.F.G.: The rheology of fresh cement and concrete — a review. In: Grieve, G., Owens, G. (eds.) Proc. of the 11th International Congress on the Chemistry of Cement: Cement's Contribution to the Development in the 21st Century, Durban, May 11-16, pp. 11–16 (2003)

[15] Wallevik, J.E.: Rheological properties of cement paste: thixotropic behavior and structural breakdown. Cement and Concrete Research 39, 14–29 (2009)

[16] Farris, R.J.: Prediction of the viscosity of multi modal suspensions from viscosity dada. Transactions of the Society of the Rheology 12(2), 281–301 (1968)

[17] Flatt, R.: Towards a prediction of superplasticized concrete rheology. Materials and Structures 27(269), 289–300 (2004)

[18] Toutou, Z., Roussel, N.: Multi scale experimental study of concrete rheology: from water scale to gravel scale. Materials and Structures 39(2), 189–199 (2006)

[19] Krieger, I.M., Dougherty, T.J.: A mechanism for non-Newtonian flow in suspensions of rigid spheres. Trans. Soc. Rheol. 3, 137–152 (1959)

[20] Geiker, M.R., Brandl, M., Thrane, L.N., Nielsen, L.F.: On the effect of coarse aggregates fraction and shape on the rheological properties of Self Compacting Concrete. Cement, Concrete and Aggregates 24(1), 3–6 (2002)

[21] Mahaut, F., Mokeddem, S., Chateau, X., Roussel, N., Ovarlez, G.: Effect of coarse particle volume fraction on the yield stress and thixotropy of cementitious materials. Cement and Concrete Research 38, 1276–1285 (2008)

[22] de Larrard, F.: Concrete mixture proportioning. E & FN Spon, London (1999)

[23] Krell, J.: Die Konsistenz von Zementleim, Mörtel, und Beton und ihre zeitliche Veränderung, PhD-thesis, Faculty of Civil Engineering, Rheinisch-Westfälische Technische Hochschule Aachen (1985) (in German)

[24] Maeyama, A., Maruyama, K., Midorikawa, T., Sakata, N.: Characterization of Powder for Self-Compacting Concrete. In: Ozawa, Ouchi (eds.) Second Int. Workshop on SCC, Kochi. Concrete Engineering Series, vol. (30), pp. 191–200. Japan Society of Civil Engineers (1998)

[25] Midorikawa, T., Pelova, G.I., Walraven, J.C.: Application of "The Water-layer Model" to self-compacting mortar with different size distribution of fine aggregate. In: Ozawa, Ouchi (eds.) Second Int. Symposium on SCC, pp. 237–246. University of Tokyo, COMS Engineering Publication (2001)

[26] Maruya, E., Osaki, M., Igarashi, H.: Relationship between rheological constant of cement paste and fluidity of high-fluidity concrete. Journal of Advanced Concrete Technology 4(2), 251–257 (2006)

[27] Perdersen, B., Smeplass, S.: The relationship between the rheological properties of SCC and the corresponding matrix phase. In: Proceedings of the 3rd International RILEM Symposium on Self-Compacting Concrete (RILEM PRO33, 2003), Reykjavik, Iceland, pp. 23–31 (2003)

[28] Wallevik, J.E.: Relationship between the Bingham parameters and slump. Cement and Concrete Research 36, 1214–1221 (2006)

[29] Wallevik, J.E.: Rheology of Particle Suspensions - Fresh Concrete, Mortar and Cement Paste with Various Types of Lignosulfonates (Ph.D.-thesis); Department of Structural Engineering, The Norwegian University of Science and Technology (2003) ISBN 82-471-5566-4, ISSN 0809-103X

[30] Yammine, J., Chaouche, M., Guerinet, M., Moranville, M., Roussel, N.: From ordinary rheology concrete to self compacting concrete: a transition between frictional and hydrodynamic interactions. Cement and Concrete Research 38, 890–896 (2008)
[31] Onoda, G.Y., Liniger, E.G.: Random loose packings of uniform spheres and the dilatancy onset. Phys. Rev. Lett. 64, 2727–2730 (1990)
[32] Mansoutre, S., Colombet, P., Van Damme, H.: Water retention and granular rheological behaviour of fresh C3S paste as function of concentration. Cement and Concrete Research 29, 1441–1453 (1999)
[33] Lootens, D., Hébraud, P., Lécolier, E., Van Damme, H.: Gelation, Shear-Thinning and Shear-Thickening in Cement Slurries. Oil & Gas Science and Technology – Rev. IFP 59(1), 31–40 (2004)
[34] Ovarlez, G., Bertrand, F., Rodts, S.: Local determination of the constitutive law of a dense suspension of noncolloidal particles through magnetic resonance imaging. J. Rheol. 50, 259–292 (2006)
[35] Oh, S.G., Noguchi, T., Tomosawa, F.: Towards mix design for rheology of self-compacting concrete. In: Skarendahl, Petersson (eds.) First Int. Symposium on SCC, Stockholm, pp. 361–372. RILEM Publications PRO 7, Cachan (1999)
[36] Denis, A., Attar, A., Breysse, D., Chauvin, J.J.: Effect of coarse aggregate on the workability of sandcrete. Cem. Concr. Res. 32(5), 701–706 (2002)
[37] Chanvillard, G., Basuyaux, O.: Une méthode de formulation des bétons de sable à maniabilité et résistance fixées. Bulletin de Liaison des Laboratoires des Ponts et Chaussées, N° 205, 49–63 (1996)
[38] Garboczi, E.J., Bentz, D.P.: Analytical formulas for interfacial transition zone properties. Advanced Cement-Based Materials 6, 99–108 (1997)
[39] Saak, A.W., Jenning, H., Shah, S.: New methodology for designing self-compacting concrete. ACI Materials Journal 98(6), 429–439 (2001)
[40] Roussel, N.: A theoretical frame to study stability of fresh concrete. Materials and Structures 39(1), 81–91 (2006)
[41] Cussigh, F., Sonebi, M., De Schutter, G.: Project testing SCC-segregation test methods. In: Proceedings of the 3rd International RILEM Symposium on Self-Compacting Concrete (RILEM PRO33, 2003), Reykjavik, Iceland, pp. 311–322 (2003)
[42] Daczko, J.A.: A comparison of passing ability test methods for self consolidating concrete. In: Proceedings of the 3rd International RILEM Symposium on Self-Compacting Concrete (RILEM PRO33, 2003), Reykjavik, Iceland, pp. 335–344 (2003)
[43] Tregger, N., Ferrara, L., Shah, S.P.: Predicting dynamic segregation resistance of self-consolidating concrete from the slump-flow test. Journal ASTM 7(1) (January 2010)
[44] Wallevik, O.: Rheology – a scientific approach to develop self-compacting concrete. In: Proceedings of the 3rd International RILEM Symposium on Self-Compacting Concrete (RILEM PRO33, 2003), Reykjavik, Iceland, pp. 23–31 (2003)
[45] Nguyen, T.L.H., Roussel, N., Coussot, P.: Correlation between L-box test and rheological parameters of an homogeneous yield stress fluid. Cem. Concr. Res. 36(10), 1789–1796 (2006)
[46] Tam, C.T., Shein, A.M.M., Ong, K.C.G., Chay, C.Y.: Modified J-ring approach for assessing passing ability of SCC. In: Proceedings of SCC 2005. Published by Hanley Wood (2005)

[47] Ng, I.Y.T., Wong, H.H.C., Kwan, A.K.H.: Passing ability and segregation stability of self-consolidating concrete with different aggregate proportions. Magazine of Concrete Research 58(6), 447–457 (2006)
[48] Roussel, N., Nguyen, T.L.H., Coussot, P.: General probabilistic approach of filtration process. Physical Review Letter 98(11), 114502 (2007)
[49] Grünewald, S., Walraven, J.C., Emborg, M., Carlswärd, J., Hedin, C.: Evaluation of test methods for filling ability of SCC. In: 2005 SCC Conference, China (2005)
[50] Kooiman, A.G.: Modeling steel fiber reinforced concrete for structural design. PhD thesis, Stevin Laboratory, Delft University of Technology (2000)
[51] Markovic, I.: High-performance hybrid-fiber concrete – development and utilisation. PhD thesis, Department of Underground Infrastructure, Delft University of Technology (2006)
[52] Stähli, P., van Mier, J.G.M.: Manufacturing, fiber anisotropy and fracture of hybrid fiber concrete. Engineering Fracture Mechanics 74(1-2), 223–242 (2007)
[53] Folgar, F., Tucker, C.L.: Orientation Behavior of Fibers in Concentrated Suspensions. Journal of Reinforced Plastics and Composites 3, 98–119 (1984)
[54] Ferrara, L., Ozyurt, N., di Prisco, M.: High mechanical performance of fiber reinforced cementitious composites: the role of "casting-flow" induced fiber orientation. Accepted for publication in Materials and Structures (January 5, 2010)
[55] Jeffrey, G.: The motion of ellipsoid particles immersed in a viscous fluid. Proc. R. Soc. London A102, 161 (1923)
[56] Roussel, N.: A thixotropy model for fresh fluid concretes: theory, validation and applications. Cement and Concrete Research 36(10), 1797–1806 (2006)
[57] Jarny, S., Roussel, N., Rodts, S., Bertrand, F., Le Roy, R., Coussot, P.: Rheological behavior of cement pastes from MRI Velocimetry. Cement and Concrete Research 35, 1873–1881 (2005)
[58] Roussel, N., Cussigh, F.: Distinct-layer casting of SCC: the mechanical consequences of thixotropy. Cement and Concrete Research 38, 624–632 (2008)
[59] Libessart, L.: Influence des composées chimiques des agents de démoulage sur l'interface coffrage/béton – Impact sur l'esthétique des parements en béton. Thesis, Artois University (2006) (in French)

Chapter 2
Computational Fluid Dynamics

Lars Thrane[1,*], Ana Bras[2], Paul Bakker[3], Wolfgang Brameshuber[4], Bogdan Cazacliu[5], Liberato Ferrara[6], Dimitri Feys[7], Mette Geiker[8], Annika Gram[9], Steffen Grünewald[10], Samir Mokeddem[5], Nicolas Roquet[5], Nicolas Roussel[5], Surendra Shah[11], Nathan Tregger[11], Stephan Uebachs[4], Frederick Van Waarde[10], and Jon Elvar Wallevik[12]

[1] DTI, Danish Technological Institute, Denmark
[2] Universidade Nova de Lisboa, Portugal
[3] Van Hattum en Blankenvoort, The Netherlands
[4] IBAC RWTH, Institut für Bauforschung Aachen, Germany
[5] IFSTTAR, France
[6] Politecnico di Milano, Italy
[7] University of Sherbrooke, Canada
[8] DTU, Danmarks Tekniske Universitet, Denmark
[9] CBI, Swedish Cement and Concrete Institute, Sweden
[10] TU Delft, Technische Universiteit, The Netherlands
[11] Northwestern University, USA
[12] Innovation Center Iceland, ICI Rheocenter, Iceland

2.1 Introduction to Computational Fluid Dynamics

In many industries where the manufacturing process involves casting of a fluid or a suspension, it is essential to have tools predicting the flow and the possible subsequent hardening process for mould design and process planning. Trial and error may be applied to optimize the casting process when a large number of small samples are produced. However, when casting concrete, especially in-situ, trial and error is rarely an option and full control of the casting process is important. Otherwise, problems such as incomplete form filling, segregation, blocking, poor encapsulation of reinforcement, poor surface finish, and cold joints may be the outcome.

The simplest assumption about the casting of a fluid or a suspension is to consider it as a Newtonian fluid subjected to laminar flow (fluid flow in parallel layers with no disruption between the layers). Simple fluids may be air and water. The aim is to solve the governing equations of flow (equations of momentum, continuity and energy) for properties such as velocity, pressure, density, and temperature. When the flow problem is simple, it is possible to obtain an analytical solution of the problem. This includes *e.g.* pipe flow relevant to pumping and

[*] Corresponding author.

N. Roussel and A. Gram (eds.), *Simulation of Fresh Concrete Flow*,
RILEM State-of-the-Art Reports 15,
DOI: 10.1007/978-94-017-8884-7_2, © RILEM 2014

so-called stoppage flow used to estimate the final shape of a surface *e.g.* in test methods. However, only few processes can be considered as “simple enough” in the field of civil engineering to allow for a purely analytical description. When dealing with more complex flows, numerical solutions are needed, in particular for transient (change over time) flows with a free surface. Computational Fluid Dynamics (CFD) is the part of fluid mechanics that refers to the use of numerical methods and algorithms to solve and analyze fluid flow problems. Besides laminar Newtonian flows, it covers topics such as turbulence flow, coupled flow problems, multiphase flows and Non-Newtonian flows [1-4]. Turbulence flow deals with disordered flow patterns *e.g.* slip stream behind cars, airplanes, and ships. Coupled flow problems refer to problems, in which some state equations are solved simultaneously with the governing equations to describe species concentration, chemical reactions and heat transfer. Multi-phase flows involve liquid/gas, solid/gas and liquid/solid interactions. Non-Newtonian flow refers to fluids with a non-linear relation between shear stress and shear rate [3, 4].

One of the first papers on a practical three dimensional problems was published in 1967 [5]. Today, CFD is being applied in many different industries. For instance, CFD has revolutionized the process of aerodynamic design and joined the wind tunnel and flight test as primary tools of the trade [6]. Also CFD has been used to predict moulding of thermoplastics and steel (*cf.* Fig. 2.1).

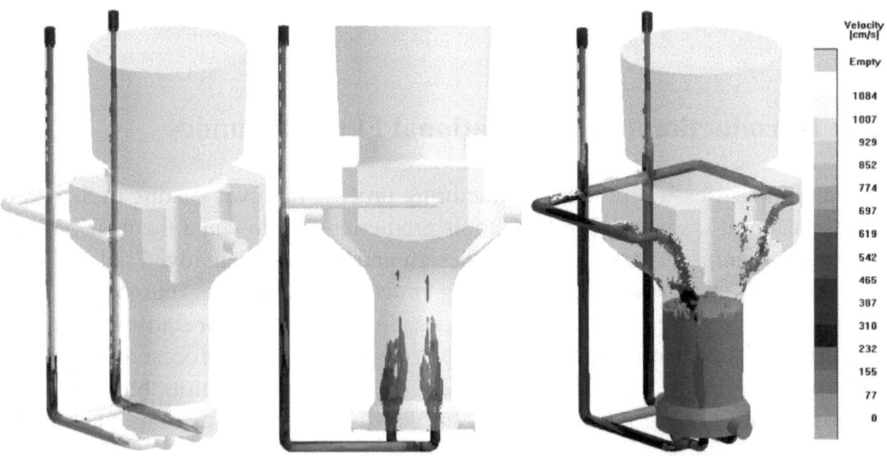

Fig. 2.1 Flow pattern at three different points in time during mould filling simulated with MAGMASOFT. Note the highly turbulent and uncontrolled flow which is very harmful for the casting quality (inclusion of oxides, *etc.*) [7].

Under certain conditions that will be discussed below, cement based materials such as concrete may, in a similar way, be treated as a fluid and simulated within the framework of CFD. The main motivations for investigating CFD in relation to flow of cement based materials are:

- Simulations can support decision making and help bridge the gap between empirical based experience and scientific based solutions to concrete flow. Thus, it supports the transfer of a rheological approach to concrete casting from R&D into practical application.
- The rheological properties, which are physical properties that can be measured, are direct inputs to the simulation. Thus, simulations may be a tool to specify the rheological properties for a specific application, which are important inputs to the mix design process and quality control procedures. In practice, at production facilities and at the job site, flow properties are measured using the simple standard test methods. Simulations of the standard test methods improve understanding and interpretation of these tests results.
- A number of materials behaviour laws can be applied. Non-Newtonian Bingham constitutive law is most often applied to cement based materials but even more sophisticated material models related to structural build up (thixotropy) have been proposed and applied [8, 9].
- CFD gives an insight into flow patterns that are difficult, expensive or impossible to study using experimental techniques.
- Compared to other methods, CFD is the fastest way to simulate flows. However, computational capacity is still an issue and a limiting factor especially for problems over large scales of time and length [10].
- The main disadvantage of the technique is that solid particles are excluded from the simulation. It is therefore not possible to directly simulate heterogeneous phenomena such as segregation and blocking.

The aim of this chapter is to give an overview of the work carried out on simulation of flow of cement-based materials using the CFD approach. This includes governing equations, constitutive equations, analytical and numerical solutions, and examples showing simulations of testing, mixing and castings. It will be shown that promising and pioneering work has already been carried out. The approach is however far from being standard in the concrete industry and clearly does not yet hold all the answers to relevant questions. In Chapter 5, future perspectives of CFD simulations will therefore be discussed.

2.2 Material Behaviour Law

2.2.1 Governing Equations

The three main equations in fluid mechanics are derived from conservation of momentum (equation of motion), mass (continuity) and energy. The solution to a fluid dynamics problem typically involves the computation of properties of the fluid, such as velocity, pressure, density, and temperature, as functions of space and time.

The equation of motion represents the primary law used when simulating the flow of cement based material. Its philosophy is Newton's second law "*m dv*/*dt* = *F*". For Newtonian fluid, the governing equation is often designated as Navier-Stokes. However, for non-Newtonian fluid, such as cement based materials, the governing equation is more complex and is named the *Cauchy equation of motion,* given by [1, 2]

$$\rho\left(\frac{\partial \mathbf{v}}{\partial t}+\mathbf{v}\cdot\nabla\mathbf{v}\right)=\nabla\cdot\boldsymbol{\sigma}+\rho\,\mathbf{g} \tag{1}$$

The term $\mathbf{v}$ is the 4 dimensional (time being one dimension) and 3 directional velocity, ρ is the density, t is the time, $\mathbf{g}$ is the gravity and $\boldsymbol{\sigma}$ is the (total) stress tensor [1, 13]. The stress tensor is given by

$$\boldsymbol{\sigma} = -\,p\,\mathbf{I} + \mathbf{T} \tag{2}$$

In general and as shown above, the stress tensor is decomposed into an isotropic pressure term p and an extra stress contribution $\mathbf{T}$, commonly known as the extra stress tensor [11]. $\mathbf{I}$ is the unit dyadic. Unfortunately, the above use of tensorial notation is most necessary and makes the use of CFD for cement based material challenging from a mathematical point of view. A good introductory text about tensor calculus with applications is given in [12].

In CFD, the continuity equation is often used to calculate pressure correction while solving Eq. (1). It writes:

$$\frac{\partial \rho}{\partial t} = -(\nabla\cdot\rho\mathbf{v}) \tag{3}$$

For liquids, which are much denser than gasses, the incompressibility constraint is often assumed, and the above equation can then be written as:

$$(\nabla\cdot\mathbf{v}) = 0 \tag{4}$$

The last of the governing equations is the energy equation, which is used to determine possible temperature variations. However, during flow of cement based materials, isothermal conditions are often assumed and the energy equation is often omitted.

2.2.2 Constitutive Equations – Generalised Newtonian Model

When using CFD for cement based material, the constitutive equation usually consists of the Generalised Newtonian Model, or in short GNM. The GNM is given by [12, 13]

$$\mathbf{T} = \eta(\mathrm{II}_d)\,\mathbf{d} \tag{5}$$

where η is the apparent viscosity (also known as shear viscosity), $\mathbf{d} = \nabla \mathbf{v} + (\nabla \mathbf{v})^T$ is the rate-of-strain tensor, while term $II_\mathbf{d}$ is the second invariant of the rate-of-strain tensor, given by [13]

$$II_\mathbf{d} = -\,[\mathrm{tr}(\mathbf{d})\,\mathrm{tr}(\mathbf{d}) - \mathbf{d}{:}\mathbf{d}]/2 \tag{6}$$

This equation represents the rate of shear intensity for three-dimensional flow, sometimes referred to as the shear rate. That is, $II_\mathbf{d} = \dot{\gamma}^2$, where $\dot{\gamma}$ represents the shear rate [13-15]. An example of apparent viscosity $\eta(II_d)$ used in Eq. (5) would of the Newtonian fluid, in which the apparent viscosity is constant (or more precisely, only dependent on thermodynamic properties like temperature [4]). Others are of the well-known Bingham and Herschel-Bulkley model given by Eqs. (7) and (8), respectively.

$$\eta = \mu + \frac{\tau_0}{\sqrt{II_\mathbf{d}}} \tag{7}$$

$$\eta = k \cdot \left(\sqrt{II_\mathbf{d}}\right)^{n-1} + \frac{\tau_0}{\sqrt{II_\mathbf{d}}} \tag{8}$$

The terms μ and τ_0 are the well known plastic viscosity and yield stress, while the terms k and n are usually referred to as consistency (or flow coefficient) and power law exponent, respectively. Eqs. (8) and (9) represents so-called viscoplastic materials. Such materials are characterized by a yield stress τ_0 that must be exceeded before significant deformation can occur. Examples of more sophisticated material models of thixotropic nature are for example presented in [8, 9, 16].

2.3 Solving a Fluid Problem

2.3.1 Global Analysis

For any flow problem, the first thing to do is to analyse and define the problem to be solved by answering the following non exhaustive list of questions:

- Is the flow laminar or turbulent? Most flow situations related to cementitious materials may be assumed to belong to the laminar flow regime. The dimensionless Reynolds number indicates the risk of turbulence (*cf.* Chapter 1).
- Is the problem steady state or unsteady (time dependent)? Steady state flow situations may occur in viscometer testing and pipe flows, whereas the evolution of flow is typically of interest in standard test methods and castings.
- Material behaviour law? Section 0 presented the concept of the generalised Newtonian model to model *e.g.* a Bingham material. However, structural build may also be relevant and theories have been proposed by Wallevik [9] and Roussel [8, 16].

- Is surface tension relevant? If the flow problems involves *e.g.* a small layer thickness of material the flow solution may need to take into account surface tension, see Roussel [17]. Surface tension is often neglected in civil engineering flow studies but may sometimes be the driving force of the flow or of its stoppage. It is the case when low plastic yield value and low viscosity materials such as cement grouts are studied. For these low concentration suspensions, the surface tension effect may be of the same order as the yield stresses (1–2 Pa).
- Does the flow problem involve a free surface? For concrete, this is typically the case in "filling" scenarios where the free surface is the interface between concrete and air. The deformation experienced by the free surface is often large and/or irregular.
- Boundary conditions? The boundary conditions may include filling conditions, *e.g.* a pressure boundary or inlet velocity condition, and fluid/solid interface conditions (no slip), moving boundaries *etc.*
- Assumptions to simplify the flow problem and reduce computation time? For instance, the energy equation is often excluded as isothermal flow conditions are assumed. The convective term may be excluded from the equation of motion when viscous forces dominate over inertia. Identified planes of symmetries may reduce the computational domain.

Before starting numerical simulations, and consequently approximating the final solution, one could verify whether an analytical solution can be found for the flow problem which needs to be solved. Most flow problems are impossible to solve analytically due to the complexity and interconnections between the different equations. However, if sufficient simplifications can be made to the set of equations, an analytical solution might be available. Secondly, analytical solutions can also be seen as a verification tool for the simulations made. A significant deviation between the numerical solution and the analytical solution (for exactly the same flow problem) might indicate a problem in the numerical code.

2.3.2 Dimensional Analysis of Concrete Flows

Before turning on the computer and starting coding some complex numerical simulations, one should always try to identify the order of magnitude of the problem parameters. In some simple cases, this approach can give enough information on the flow to completely avoid any computational analysis. In this section, we do not resolve or address detailed processing problems. We only focus on identifying the physical parameters which govern the concrete flows patterns.

2.3.2.1 Dimensional Analysis of Slump and Slump Flow Tests

In the case of slump or slump flow test, gravity generates stresses in the sample when the mould is lifted. The order of magnitude of this stress is $\rho g(H_0\text{-}S)$ where H_0 is the initial height of the cone (30 cm), S is the slump and ρ, the density of concrete, is of the order of 2500 kg/m^3. Flow stops when this stress is equal to yield stress. It can be concluded from this simple dimensional analysis that concretes with no slump have a yield stress higher than several thousands of Pa and that self compacting concretes with slumps higher than 25 have yield stresses lower than 150 Pa.

2.3.2.2 Standard Shear Flows in Industrial Practice

As already discussed, one of the specific features of cementitious materials comes from the fact that they flow only if they are applied a stress higher than a critical value called yield stress [18-21]. This yield stress has several important consequences on the flow pattern. It is possible to predict some of them by using a simple dimensional analysis of the flow pattern. First, there may exist in the flow some areas which are not submitted to any deformations as the stress in these zones is lower than the yield stress. In these so-called "plug-flow" areas (if they are carried by the rest of the material) or "dead zones" (if they do not move at all), all deformation rates of the fluid are equal to zero. Second, when a macroscopic apparent shear rate is applied to the material, as some parts of the material are not flowing, shear concentrates in sheared zones, in which it can reach values far higher than the apparent shear rate applied to the material. It can be noted that, in many industrial flows, these high shear rates zones are located at the interface with the mold.

Most industrial flows in practice can be divided in two categories: free surface flows (*i.e.* above one interface) and confined flows (*i.e.* between two walls) as shown in Fig. 2.2. Most horizontal applications such as slab casting belong to the first category whereas most vertical applications such as wall casting belong to the second one. We will call here e the characteristic size of the flow. We define e as the thickness of the flowing concrete layer in the case of free surface flows and as half the distance between the two walls in the case of confined flows.

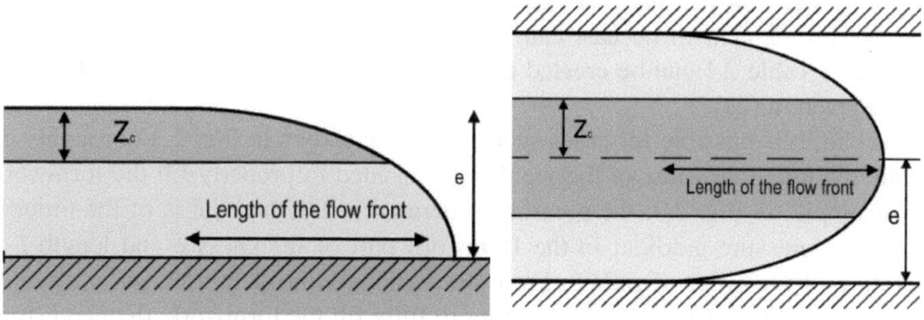

Fig. 2.2 The two categories of flow in industrial practice. (left) free surface flow referred as slab casting (right) confined flow referred here as wall casting.

Gravity is, in most casting processes in the construction industry, the "engine" of the flow (*i.e.* concrete is seldom injected under external pressure). The pressure gradient generated by gravity in the material is equal to zero in the zones where the material surface level is horizontal and ρgH for purely vertical flows where ρ, the density of concrete, is of the order of 2500 kg/m^3. The pressure gradient depends on the free surface shape of the material during casting, which itself depends on the casting process and rheological behaviour. It is therefore very difficult to identify a typical value. However, as a first approximation, we will consider here that the typical thickness of a concrete flowing layer is 10 cm and that the typical length of the flow front is of the order of 20 cm in the case of Ordinary Rheology Concrete (ORC) and 50 cm in the case of Self Compacting Concretes (SCC). The pressure gradient is then of the order of $\rho g/2$ in the case of ORC and $\rho g/5$ in the case of SCC.

Conservation of momentum imposes that the pressure gradient equals the shear stress gradient at any points in these shearing flows. As a consequence, the shear stress in the two configurations in Fig. 2.2 shall linearly decrease with the distance from the wall (up to the free surface or up the symmetry plan). There shall therefore exists a zone of thickness z_c (or 2 z_c in the case of confined flows) in which the stress is lower than the yield stress. The extent of z_c shall be of the order of $2\tau_0/\rho g$ in the case of ORC or $5\tau_0/\rho g$ in the case of SCC. If z_c is higher than e, no flow occurs at all and this means that the concrete is too stiff for the casting of the element. Most of the time, fortunately, z_c is lower than e and concrete is able to fill the structural element. The role of vibration, which is able to reduce the yield stress value [20], is then limited to the extraction of excess air bubbles.

2.3.2.3 Filling Prediction

First, it shall be kept in mind that the highest pressure gradient that gravity can create in the material is ρg. From a dimensional point of view, the smallest characteristic dimension (distance between formwork walls or distance between steel bars) of an element to be cast can not be smaller than $\tau_0/\rho g$. From this simple statement, Table 2.1 can be created using the correlation between yield stress and slump given in [21].

Second, it is possible for cases such as the one shown in Fig. 2.3 to roughly estimate the maximal value of the yield stress needed to properly fill the formwork. For example, in Fig. 2.3, the pressure generated in the material is of the order of ρgH. The pressure gradient in the formwork part of thickness e and length L is therefore of the order of $\rho gH/L$. From dimensional considerations, shear stress in this zone is of the order $\rho gHe/L$. In order to fully fill the formwork, the yield stress of the chosen material must be lower than this value.

Table 2.1 Minimal characteristic size of the element to be cast as a function of either slump or yield stress (effect of vibration is not taken into account)

slump (cm)	Yield stress (Pa)	Minimal characteristic size of the element to be cast (cm)
0	3500	15
5	3000	13
10	2000	8
15	1500	6
20	750	3
25	100	0.4 (lower than the size of aggregates)

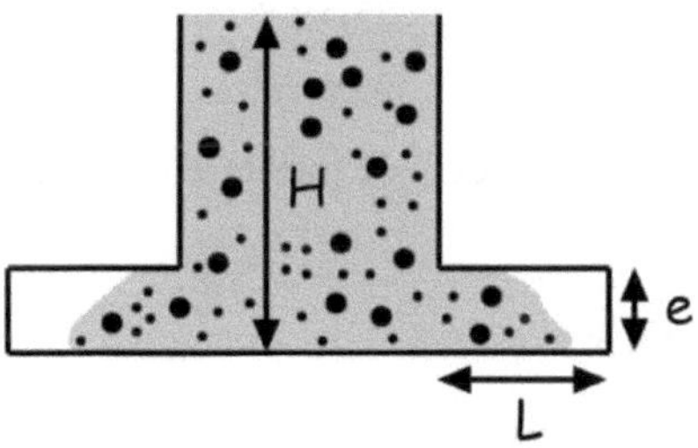

Fig. 2.3 Simple element to be cast

2.4 Ananlytical Solutions

2.4.1 Free Surface Flow

2.4.1.1 Slump and Slump Flow

The best example of a free surface flow test is the slump/slump flow test. The slump/slump flow test is also the most simulated test, as it can provide an analytical verification of the yield stress. Roussel and Coussot [22] described the relation between slump/slump flow and yield stress from an analytical point of view for two situations: $H >> R$ and $H << R$ (H is the height; R is the radius of the sample). In these two situations, the governing flow equations can be significantly simplified in order to obtain an analytical solution.

Shearing Flow: *H* << *R*

In case of $H << R$, which corresponds to the slump flow regime, the three-dimensional strain rate tensor simplifies to a one-dimensional equation, as the fluid height is much smaller than its radius, the radial velocity is much larger than the vertical velocity ($v_r >> v_z$) and the variation of flow characteristics is much larger in the vertical direction than in the radial direction ($d/dh >> d/dr$). In this case, the yielding criterion is 1-dimensional and flow stoppage will be achieved when the shear stress is equal or lower than the yield stress in the entire material sample. Following the procedure presented in [22] the final shape in the case of stoppage flow (fluid is at rest) can be calculated as:

$$h(r) = \left(\frac{2\tau_0 (R - r)}{\rho g} \right)^{1/2} \tag{9}$$

where h is the height, r the radius, R the final spreading radius, τ_0 the yield stress, ρ the density and g the gravity. Knowing the sample volume Ω, the following equation delivers the relation between the yield stress and the spreading radius of the sample:

$$\tau_0 = \frac{25 \rho g \Omega^2}{32 \pi^2 R^5} \tag{10}$$

Note that, for concrete, the effects of inertia and surface tension are negligible in the final solution [22].

Elongational Flow: *H* >> *R*

In the case $H >> R$, indicating very low slump concrete, the main stress variation occurs in vertical direction due to gravity. In this case, there is no significant shearing flow and the relation between slump (z_0) and yield stress is obtained by the Von Mises yielding criterion:

$$\tau_0 = \frac{\rho g (H - z_0)}{\sqrt{3}} \tag{11}$$

Intermediate Cases

In the intermediate cases, when H and R are of the same order of magnitude, analytical solutions are not available, as the above simplifications are no longer valid. Approximate solutions can be obtained by other simplifications and assumptions or by numerical simulations. The pure elongational flow ($H >> R$) is only obtained for slumps very close to zero as no flow is assumed to occur, but the equation obtained in the pure shearing flow ($H << R$) is sufficiently reliable in the slump flow regime ($H/R < 0.2$). Fig. 2.4 shows a plot of slump versus yield stress, showing the validity of the two analytical solutions for $H << R$ (left) and $H >> R$ (right) [22].

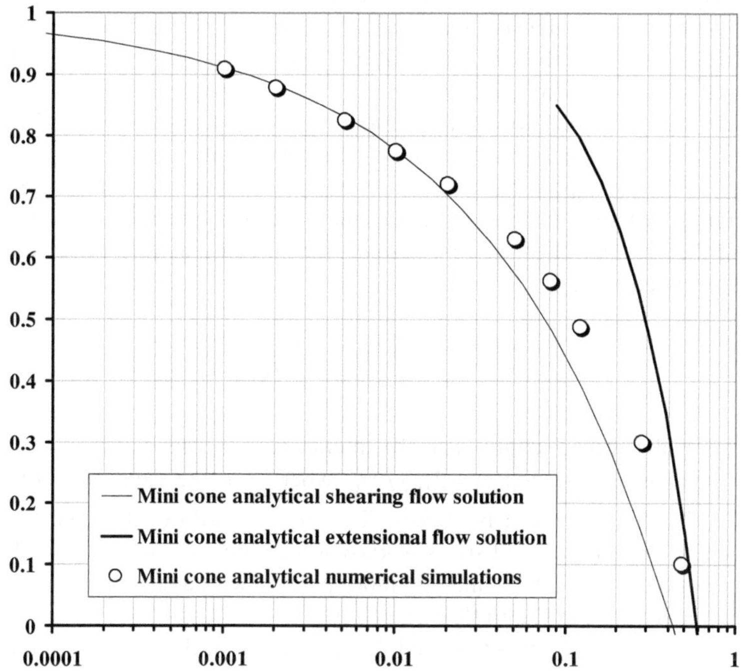

Fig. 2.4 Dimensionless slump versus dimensionless yield stress. This figure shows the validity of the obtained analytical solutions in the two asymptotic regions (redrawn from [22]).

One should note that in case of fluid concrete, the dimension of the largest particle becomes of the same order of magnitude as the smallest dimension during flow (the height of the final slump flow). Some local heterogeneity may then play a strong role on the resulting behaviour (see Chapter 1). As all analytical calculations assume a homogeneous suspension, the validity of the above relations between slump flow and yield stress can be questioned.

2.4.1.2 Channel Flow

The L-box is another tool to measure the yield stress of cement based materials. In contrast to the slump or slump flow, the final result of the L-box test is influenced by inertia if the gate is opened rapidly [23, 24]. Due to inertia effects, an L-box ratio (h_2/h_1) larger than 1 can be obtained, especially in case of low viscosity materials. In the extreme case, inertia might even be dominating and the material splashes out at the end of the box. When inertia cannot be neglected, an analytical solution is not straightforward as the final shape of the material is not only determined by its density and yield stress, but also by the plastic viscosity and the lifting speed of the gate. In order to avoid the influence of inertia, the gate should be lifted slowly (during at least 10 seconds). In this case, the final shape of the material only depends on its density and yield stress [24]. To differentiate between the standard L-box test and the one with a slow lifting of the gate, we will call in the following the last one, channel flow.

Considering a channel flow without steel bars, the stress tensor can be simplified to the shear stress near the bottom and the walls of the channel. Following the procedure in [24] it is possible to write an analytical solution in the two cases, *a*) the concrete is reaching the end of the channel, and *b*) the concrete is not reaching the end of the channel. For the first case *a),* the spread length in the channel may then be calculated using:

$$L = \frac{h_0}{A} + \frac{l_0}{2A} \ln\left(\frac{l_0}{l_0 + 2h_0}\right) \tag{12}$$

Where L is the spread length, h_0 is the thickness of the deposit at $x = 0$, l_0 is the width of the channel (m), and $A = 2\tau_0 / \rho g l_0$ where τ_0 is the yield stress of the material, ρ is the density and g is gravity. This relation allows the prediction of L (spread length) when the yield stress is known for a given concrete [24]. The estimation of concrete yield stress may be more precise with a sufficiently long channel test, compared to the slump flow test [25].

For the second case *b*), it is assumed that the height of the material at the end of the channel is much smaller than the channel width. In this case, the solution can be rewritten as:

$$L_0 = \frac{h_1 - h_2}{A} + \frac{l_0}{2A} \ln\left(\frac{l_0 + 2h_2}{l_0 + 2h_1}\right) \tag{13}$$

With L_0 the total length of the channel (m) and h_1, h_2 the respective thicknesses of the material at $x = 0$ and $x = L_0$.

By approximating the total volume of the sample $V = l_0 L_0 (h_1 + h_2)/2$, thus assuming a linear variation of the height with the length (which is fairly correct), and by introducing the dimensionless parameters $r = h_2/h_1$ and $V' = V/(l_0^2 L_0)$, the following dimensionless equation, expressing the relationship between the channel flow ratio r and the yield stress, can be obtained.

$$4V' \frac{1 - r}{1 + r} - \ln\left(\frac{1 + r + 4V'}{1 + r + 4rV'}\right) = \frac{2AL_0}{l_0} \tag{14}$$

Adding steel bars should not significantly influence the final shape of the material, under the condition the material stays homogeneous. This can be doubted if the concrete segregates or blocks during the flow between the rebars [24].

2.4.2 Confined Flow

In case of confined flow, analytical solutions are available for the flow of a liquid in a cylindrical pipe. For Newtonian liquids (like water), the Poiseuille equation [26] describes the relation between the total pressure (p in Pa) and the flow rate (Q in m^3/h), as a function of the length of the pipe (L in m), the viscosity of the liquid (μ in Pas) and the diameter of the pipe (D in m).

$$p = \frac{128 \mu L Q}{\pi D^4} \tag{15}$$

The Poiseuille formula is only valid if [26]:

- The flow is fully developed and isothermal
- The liquid is incompressible and homogeneous
- There are no tangential or radial flow components (1D-flow)
- The flow is steady
- There is no wall slip
- The liquid is Newtonian
- The flow is laminar

For pumping of concrete, most of the conditions can be assumed to be fulfilled, except the fact that the liquid is non-Newtonian. The Poiseuille formula can be extended as follows:

- Based on the equilibrium of forces in the pipe, the shear stress at the wall is related to the pressure loss per unit of length. Further, the shear stress decreases linearly from its maximal value at the wall to zero in the centre, no matter the rheological properties of the liquid.
- Incorporating the relationship between shear stress and shear rate in the shear stress distribution delivers the variation in shear rate, and integrating this variation over the radius delivers the velocity distribution, assuming zero velocity at the wall (no slip).
- The integration of the velocity profile over the pipe surface delivers the flow rate (Q).

In this way, a similar equation can be obtained for Bingham liquids, which is known as the Buckingham-Reiner equation [27]:

$$Q = \frac{1}{384} \frac{\pi \left(3D^4 (\Delta p)^4 + 256 \tau_0^4 L^4 - 16 \tau_0 L D^3 (\Delta p)^3\right)}{(\Delta p)^3 L \mu} \tag{16}$$

For the so-called modified Bingham model, which takes into account the shear thickening behaviour of concrete at high shear rates and was tested in [28, 29], a further extension of Eq. (16) can be obtained. For Herschel-Bulkey, due to mathematical restrictions, the analytical solution cannot be obtained for shear-thickening liquids (n>1) [28, 29]. The analytical solutions obtained for pumping of concrete, however, over-estimate the experimentally measured relationships between pressure loss and flow rate [30, 31]. This can be attributed to two different causes: slippage near the wall and/or non-homogeneity of the concrete in the pipes [30, 31]. In the latter case, the term "lubrication layer" is also employed (*cf.* Chapter 1). The three different boundary conditions (no slip, slip, and the lubrication layer) are visually represented in the following figure [32]:

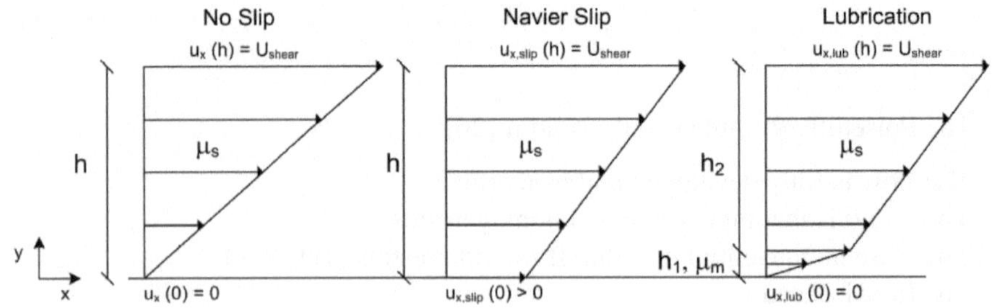

Fig. 2.5 Velocity profile in case of no-slip (left), slip (middle) and the formation of a lubrication layer (right). Figure from [32].

As mentioned before, slippage or lubrication layer certainly occur when pumping concrete through pipes. Depending on the applied shear stress and on the yield stress of the concrete, the lubrication layer (or slippage layer) takes all slippage (high yield stress, low shear stress) or shearing also occurs in the homogeneous part of the concrete (low yield stress, high shear stress). In his Ph.D, Kaplan made a distinction between those two types of flow [30].

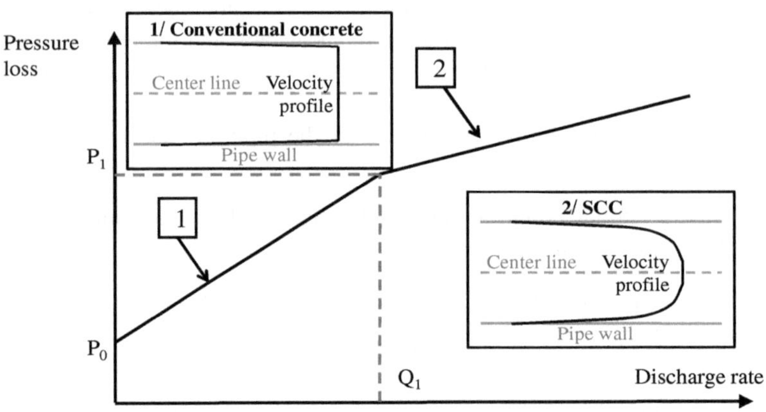

Fig. 2.6 Distinction between zone 1 and zone 2 during pumping of concrete. In zone 1, all shearing is concentrated in the lubrication layer, while in zone 2, the bulk concrete is also partly sheared. Figure after [30].

With the development of a "tribometer", which measures the relationship between stress and flow velocity in case slippage is not prohibited near the wall, the properties of the lubrication layer (yield stress (τ_{0i} in Pa) and viscous constant (η in Pas/m)) can be determined [30, 31].

In zone 1, the bulk concrete is not sheared. Consequently, only the tribological properties of the concrete influence the pressure loss – discharge rate curve by means of the following equation [30]:

$$p = \frac{2L}{R}\left(\tau_{0i} + \frac{Q}{3600\pi R^2}\eta\right) \tag{17}$$

In zone 2, the bulk concrete is partly sheared and the pressure loss – discharge rate curve is determined by a combination of rheological and tribological properties. However, in this case, the determination of the tribological properties is not obvious as the thickness of the lubrication layer is difficult to measure and the distinction between zones 1 and 2 is difficult to make [30].

$$p = \frac{2L}{R}\left(\tau_{0i} + \frac{\frac{Q}{3600\pi R^2} - \frac{R}{4\mu}\tau_{0i} + \frac{R}{3\mu}\tau_0}{1 + \frac{R}{4\mu}\eta}\eta\right) \tag{18}$$

Note that for the equations including the lubrication layer, Bingham behaviour for both the concrete and the lubrication layer has been assumed. In the work of Kaplan [30] and Chapdelaine [31], the measurement of tribological behaviour and the implementation of the parameters in the proper equations allowed a good estimation of the pumping pressures for conventional concrete (in zone 1). Only a small number of the reported tests were located in zone 2.

For pumping of concrete, as long as the exact behaviour of the concrete near the pipe wall is not understood and measured, it is questionable to make correct numerical simulations allowing for a prediction of the pumping pressure. Tribometers serve as a valuable tool to make a real case estimation, but more research on this topic is needed to further clarify the real physical phenomena involved, before starting numerical predictions of pumping pressures.

2.5 Numerical Solution

The aim of numerical solutions is to convert the partial differential equations (PDE) into a set of algebraic equations which are solved at discrete points in space and time. The three main discretization approaches to transform the PDE into a set of ordinary differential equations are the Finite Differences (FD), Finite Volumes (FV) and Finite Elements (FEM) technique. The topic of numerical simulations in fluid dynamics is very wide and covers a lot of numerical details. The following introduces just the very basic steps in finding a numerical solution to a given flow problem. For further reading and details, please take a look at *e.g.* [33-35] and also Chapter 4 on numerical errors in CFD.

One of the first steps is to decompose the geometry into cells or elements, the so-called meshing procedure. The grid may be structured or unstructured. In 2D, it may consist of triangle or quadrilateral elements whereas, in 3D, brick and tetrahedradon may form the basic elements. The computational nodes are placed close to each other in order to satisfy the continuum hypothesis. It is important to optimize the density of computational grid to obtain accurate solutions. However, it should not be over dense, which will just lead to extra computational time.

The algorithms of continuum mechanics usually make use of two classical descriptions of motion: the Lagrangian description and the Eulerian description [1, 36]. One of the main advantages of the Lagrangian approach in comparison to the Eulerian approach is the ability to model history dependent materials. Its weakness is its inability to follow large distortions of the computational domain without recourse to frequent re-meshing operations, for which reason Eulerian algorithms are widely used in fluid dynamics The Eulerian grid is fixed on the space, in which the simulated object is located. Therefore, all grid nodes and mesh cells remain spatially fixed in space and do not change with time while the materials are moving across the mesh [1]. The shape and volume of the mesh cell remain unchanged in the entire process of the computation. In the Eulerian description, large distortions in the continuum motion can be handled with relative ease using the so-called volume of fluid technique developed by Hirt and Nicols [37]. Attempts to combine the two methods have been proposed *e.g.* the arbitrary Lagrangian-Eulerian description has also been proposed [1] and, recently, a Lagrangian finite element approach for the solution of non-Newtonian fluid flow was proposed based on the so-called Particle Finite Element Method [38]. In this latter case, to better deal with free-surface flow, the fluid problem is formulated in a Lagrangian framework. To avoid excessive mesh distortion, a continuous re-meshing is introduced based on a Delaunay triangulation. One of the key features of this approach is the recovery of external and internal boundaries by means of the alpha-shape technique, which also allows the tracking of free-surface [38].

Subsequently, initial and boundary conditions are defined. Initial conditions may be the initial position of the fluid *e.g.* its position in the slump cone or vertical part of the L-box. Boundary conditions define *e.g.* the interaction between the fluid and solid walls. For instance, the velocity components of the boundary nodes may be constrained to a predefined value (Dirichlet boundary condition). For a no-slip condition, the velocity at the boundary is set to zero.

The resulting system of non-linear equations is then solved to determine the degrees of freedom, *i.e.* velocity and pressure. The equations must be solved iteratively. The methods comprise *e.g.* successive substitution, Newton-Raphson, Modified-Newton, Quasi-Newton (Broyden's update) and segregated solver. This phase often accounts for the majority of computation time. It is important to choose the solution method best suited to the problem. For time dependent problems, time also needs to be discretized. Explicit or implicit methods may be applied like *e.g.* an explicit forward Euler method or implicit backward Euler method.

One of the main difficulties in connection with fluid simulations are the yield stress behaviour of the material and its effect on the free surface displacement. Indeed, the apparent viscosity of the material is, most of the time, applied in the governing equations in order to obtain a numerical solution of the problem. However, the apparent viscosity of a yield stress fluid approaches infinity when the shear rate (or more generally in 3D the strain rate or the second invariant of the strain rates tensor) approaches zero. It is necessary to avoid this indetermination of the deformation state below the yield stress in zones where flow stops or starts, which are most of the time the zones of interest or the free surface.

2.6 Simulation of Fresh Cementitious Materials

Many researchers have carried out simulations of flow within the framework of Computational Fluid dynamics. Simulations have mostly been used to model the flow of fresh concrete during testing including standard test methods and viscometers. However, a few examples of computational modelling of mixing and full-scale castings assuming single fluid behaviour can also be found.

2.6.1 Standard Test Methods

Simulation of the standard test methods have mainly focused on three flow domains; 1) cone geometry as in the Mini-cone and Abrams cone, 2) channel flow as in the L-box, and 3) funnel flow as in the V-funnel test and Marsh cone.

Mori and Tanigawa performed some of the first simulations of fresh cementitous materials based on the so-called Viscoplastic Finite Element Method (VFEM) and the Visco-plastic Divided Element [39]. Both methods were found applicable for simulation of various test methods at reasonable computation time. VFEM was used by Kurokawa *et al.* [40] to evaluate factors affecting the slump flow of fresh concrete. The Bingham model was applied and the rheological properties were determined from the best fit to experimental results of slump flow. It was also observed that the rate at which the slump cone was lifted had an effect on the evolution of the slump, but little effect on the value of the final slump. Christensen [41] was able to enhance their slump flow simulation material model using finite elements and the software FIDAP. Later simulations of standard test methods may be found in *e.g.* Roussel [21, 22], Thrane *et al.* [23, 32, 42], Brameshuber and Uebachs [43, 44], Tregger *et al.* [45], Van Waarde *et al.* [46, 47], Gram [48] and Bras [49].

Roussel used Flow 3D® to perform 3D simulations of different slump test methods [21, 22]. An elasto-viscoplastic model was used to describe the fluid behaviour of concrete with yield stresses between 25 and 5500 Pa, assuming an incompressible and elastic solid up to the yield stress and a Bingham fluid beyond that. A no-slip condition was applied to the base. Good agreement between numerical and experimental results was obtained for the Mini-cone test and for the ASTM slump test. Good agreement was also found between numerical and analytical solutions at stoppage flow for pure extensional flow and pure shearing flow (*cf.* Fig. 2.4). It has to be noted that, in order to obtain this good quantitative agreement, Roussel but also Thrane *et al.* in [23, 42] had to implement a proper three dimensional yield criterion. Examples of two-dimensional predicted shapes are shown in Fig. 2.7 for the ASTM Abrams cone [21]. The presence of a non yielded upper zone (*i.e.* non flowing zone) usual in this type of simulation) can be noted. The calculated values of the slump confirmed the fact that slump (*i.e.* final shape) only depends on yield stress and density of the concrete.

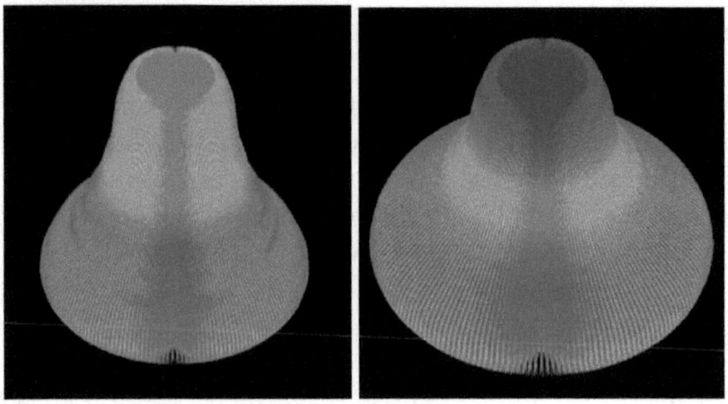

Fig. 2.7 Examples of obtained shapes for the ASTM Abrams cone (left) yield stress = 2600 Pa (right) yield stress = 2000 Pa. density = 2500 kg/m^3 for both simulations (from [21]).

Thrane *et al.* have used FIDAP® to simulate SCC flow during L-box and slump flow test assuming a Bingham behaviour [23, 32, 42]. An axi-symmetric model was applied for the slump flow test and a 3-D model was applied for half of the L-box test (symmetry plane applied). The simulations included moving boundaries for the cone in slump flow test and the gate in the L-box. For the fixed boundaries, a no-slip condition was applied. Combined simulations and experiments indicated that the transient flow in the slump flow test and the flow in an L-box with reinforcement can be simulated assuming an ideal Bingham behaviour and rheological properties were measured in a BML rheometer. The important roles of the lifting of the cone (or gate) and of the 3D model for simulating the flow in the L-box were demonstrated. The results also showed a good correlation between the simulation and experimental observations of the shape of the free surface during flow in the L-box (Fig. 2.8). Also, the effect of slip using a Navier slip condition such as the one in Fig. 2.5 was investigated. As expected, the resulting spread diameter was higher when applying a slip [32].

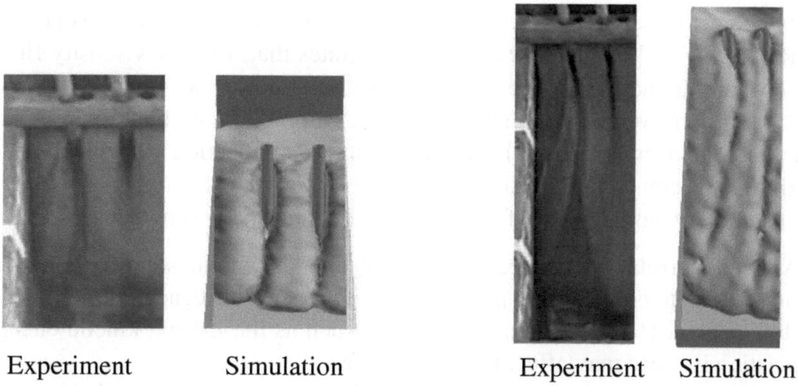

Fig. 2.8 Comparison of the free surface shape in the L-box test observed in the experiment and obtained from simulations at a flow distances of 0.35m (left) and 0.60m (right) (from [32])

Uebachs and Brameshuber [43] did also study L-box test using the software FLUENT®. They found good correlation between experiment and simulation for Newtonian fluids. Tregger *et al.* [45] applied POLYFLOW® to study the Mini-Cone flow test. Calibration of the modelling approach was performed with reference to mini-cone flow tests of cement pastes (different compositions for a wide range of yield stress and viscosity).

Van Waarde *et al.* [46, 47] applied FLUENT® to simulate the flow through a small funnel. The predictions were compared to experimental funnel tests of cement and bentonite-cement mixtures. The funnel was modelled with one quarter of the real funnel to shorten the computation time. The no-slip boundary condition was applied. A Rheolab MC1-rheometer was used to measure the yield stress and plastic viscosity as input-parameters for the numerical calculations. There was a rather good agreement between simulations and measurement of the funnel times considering the complex rheological conditions and other influencing parameters on the funnel-time such as occasional blocking, thixotropic behaviour and possible slip conditions.

Bras [49] applied FLUENT© to simulate the Marsh cone test (according to EN445). The aim was to relate the rheological parameters of some hydraulic lime grouts (for masonry injection and consolidation) to their flow time through the Marsh cone, which characterizes in a practical way the fluidity of grouts. A simple approximation using Bingham behaviour was adopted. The grouts presented various rheological characteristics: the yield stress varied from 0.34 to 2.13 Pa and plastic viscosity from 0.09 to 0.23 Pa.s. The effect of "no-slip" parameter in Marsh cone walls was analysed and it was observed that this condition led to the best prediction of the test results. A comparison of numerical and experimental results showed good correlation for the least fluid materials. It was suggested that, for the most fluid materials and short Cone nozzles, the grout flow may become turbulent when the Reynolds number was higher than 10 (*cf.* Chapter 1). In these cases, the velocity may be overestimated [50]. This conclusion was also reached in Le Roy *et al.* [51]. Roussel *et al.* [52] also states that, for low viscosity fluids because of turbulence, the cone flow time value is not a meaningful measurement and that a narrower nozzle shall be chosen. In practice, it should be emphasized that the Marsh cone has to be used within its application domain to reach its maximum efficiency as a control tool [51].

In summary, simulations of standard test methods have shown:

- A good correlation between numerical and analytical solution for stoppage flow. Results confirm that plastic viscosity has no influence on the final slump flow spread. However, in specific cases such as the L-Box, kinetic energy and plastic viscosity may affect the final shape.
- A good correlation between transient flows observed in simulations and experiments when using proper boundary conditions, in particular moving boundaries related to lifting of the cone in the slump flow test and the gate in the L-box. A reasonable agreement has been found between measured rheological parameters and the ones used in simulations. However, only limited data are available and further research is needed where measured rheological parameters are used as input parameters into simulations and compared to experimental observations of transient flow.
- A good agreement between the time evolution of the shapes of free surfaces observed in simulations and experiments.
- That the assumption of laminar flow is only valid for low Reynolds numbers as shown by Marsh cone studies of low viscosity grouts.
- That it is possible to simulate wall slip using, for instance, a Navier slip boundary condition. However, a no slip boundary condition is often applied in literature and seems to give an adequate description of the fluid-wall interface. In the case of pumping processes simulations, further research could investigate the determination of slip related parameters using tribometers [44].

2.6.2 Viscometers

The objective of these simulations is often to investigate the flow inside different viscometers in order to verify assumptions allowing for the conversion of macroscopic measurements such as torque or rotational viscosity into physical quantities such as yield stress or viscosity.

Analysis of early models of concrete rheometers was largely based on continuum approaches. Hu *et al.* [53] used a finite element method to study the BTRHEOM. The authors were able to simulate slippage or not at the walls on the rheometer container, as well as determine the uncertainty of the results. More recently, the ConTec BML Viscometer 3, ConTec Viscometer 4 following viscometers have been analyzed by computational means [54]. Additional objectives were to investigate potential new rheometers (Cone-Geometry or Parallel Plates Geometry [54]).

The simulations are based on the assumption of a visco-plastic material, flowing either under steady state or time dependent ("transient") conditions. Wallevik used a combination of several different techniques to describe the visco-plastic behaviour of the concrete. Using a freely available numerical software VVPF [55], the velocity and shear stress profiles for various viscometer configurations were simulated. The computational modelling was used for the comparison of rheometers with regard to, among others, variations in shear rate and particle migration.

2.6.3 Mixing

Contrarily to the above sections in which a homogenous material was studied, concrete mixing process consists in blending solid components (cement, sand and stones) and liquids (water, admixtures). The mixing blades generally follow complex motions, such as planetary motions. The change in microstructure and heterogeneity is fast, as the final material is often obtained in less than one minute. From a numerical simulation point of view, the number of solid particles and their poly-dispersity make a discrete approach unreachable (as well as the coupling with the fluid dispersion) without strong simplifications (*cf.* chapter 3). The difficulties to overcome belong to the following list:

1. complex changing rheology
2. heterogeneity
3. flow domain with complex geometry
4. free surface flow
5. possibly non-linear wall conditions (such as yield-strength slip phenomenon)
6. localized flow (in the neighbourhood of the blades)

CFD efficient techniques exist to deal with points 2, 3, 4 and 6. Points 1 and 5 still require however some research effort to determine the constitutive laws governing the phenomena. Accounting for the whole set 1-6 leads to very costly computations when standard (and thus generic) methods and codes are used. As an example, IFSTTAR (ex-LCPC) process modelling group decided to develop a home-made code strictly limited to mixing applications and researches in this area. In particular, accounting for new laws in points 1 and 5 require a full access to the flow solver for a possibly strong modification. Bingham and Herschel-Bulkley laws are considered as suitable constitutive models for concrete under mixing (no thixotropic effects). However, at earlier stages of mixing, additional visco-plastic-frictional-dilatant (VFD) effects are expected [57].

The following set of numerical methods were adapted, extended and combined:

7. characteristics methods [58] for time discretization, accounting for the full material derivative (*i.e.* accounting for inertial effects),

8. cartesian grid/fictitious domain approach (FD) based on low order finite element [59] for the spatial discretization (in particular : Discrete Langrange-multiplier Method, DLM [60], Immersed Finite Element, IFE [61]), in order to get robust and economic simulations without loosing a suitable accuracy.
9. AMR techniques (Adaptive Mesh Refinement) [62] allowing for local mesh refinements without loosing the benefit of FD approach,
10. non-linear non-differentiable saddle-points solvers [63], suitable for yield-stress based constitutive laws. This is a key feature as a core of the numerical solver, hopefully suitable for future rheological laws as well as non-trivial wall laws (in particular considering the slip-law in [64]).

Currently, this home-made code is able to perform 3D and 2D simulations for a Bingham material in any full (*i.e.* no free surface) planetary or annular mixer. Implementation of AMR and IFE algorithms is under construction, as well as the integration of a VFD law. Recent results in 2D are shown in Fig. 2.9.

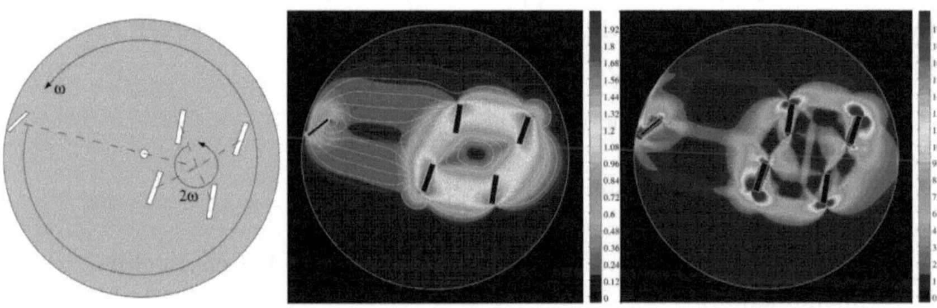

Fig. 2.9 (left) 2D model of the planetary industrial concrete mixer; (centre) velocity field at time t=10s. Colours represent velocity magnitude, yellow lines are streamlines; (right) Rate of deformation in the industrial mixer at time t=10s.

The flow of coarse grains suspension as also simulated, using a distributed Lagrange multiplier technique [60]. The suspending fluid was a Bingham paste. Fig. 2.10 represents an example of simulation of the homogenization of particles in a simple mixer. It must be noticed that particles really interact with each other and with the suspending fluid and are not passive markers. This is a first step towards the understanding of double-mixing mechanisms. The two regimes in the segregation curve (Fig. 2.11) together with pattern analysis of the cloud of particles suggests that particles transport first dominates, then a much slower mechanism of diffusion (due to dilatancy near the blades and chocks between particles) achieves the homogenization process.

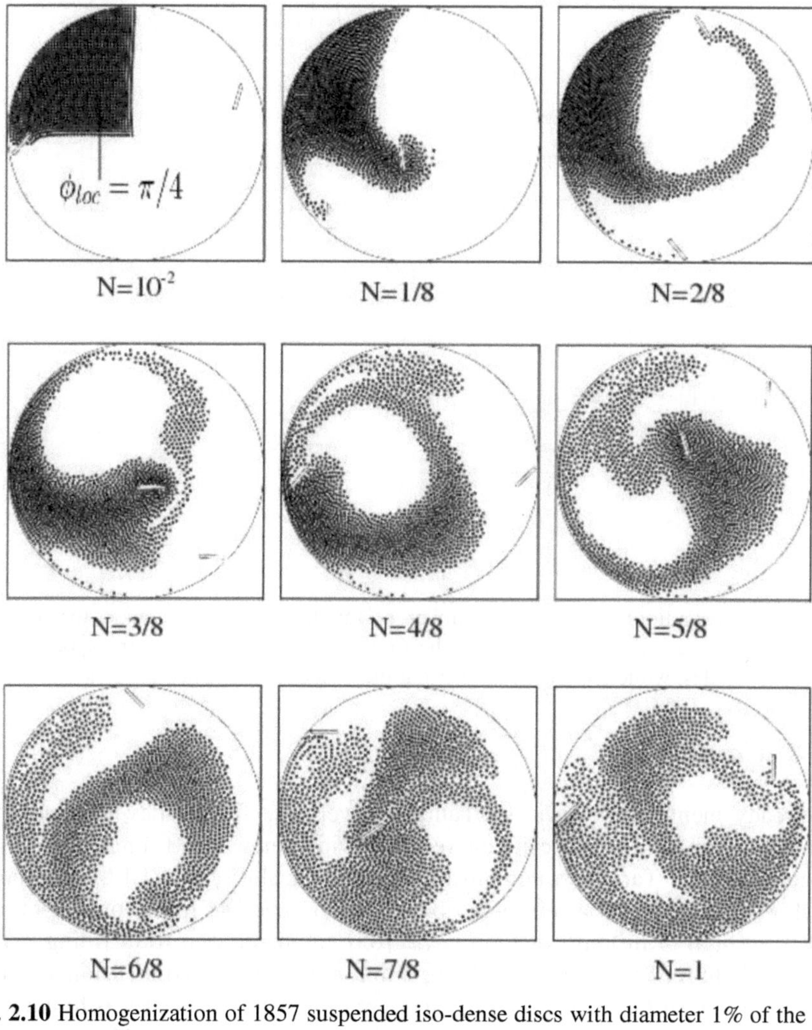

Fig. 2.10 Homogenization of 1857 suspended iso-dense discs with diameter 1% of the vessel diameter for increasing number of rotations

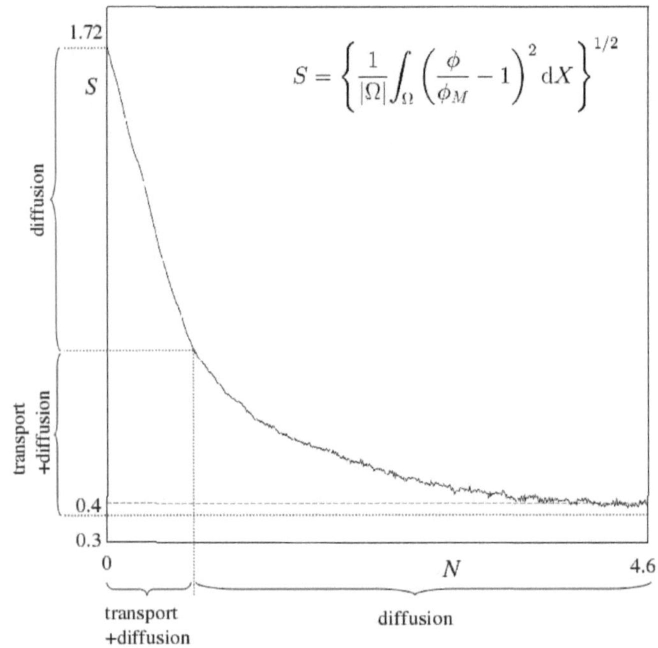

Fig. 2.11 Segregation index versus number of revolution of the scraper, for the 1857 iso-dense suspended discs. It seems that the first regime corresponds to transport of particles by the blades, while the second slower regime is due to a diffusion process.

2.6.4 Casting

As already mentioned, Mori and Tanigawa were the first to develop and apply models for simulation of fresh concrete behaviour including filling [39]. For instance, Mori and Tanigawa demonstrated the applicability of VDEM to simulate the flow of concrete in a reinforced beam section and the filling of a reinforced wall (2 m high, 3 m long). Kitaoji *et al.* [65] compared two form filling experiments without reinforcement (1 m high, 2 m long) with a two-dimensional simulation of the free surface using VDEM, and a reasonable agreement was found using $\tau_0 = 50$ Pa and $\eta_{pl} = 800$ Pa.s in the first and $\tau_0 = 300$ Pa and $\eta_{pl} = 1100$ Pa.s in the second form filling. Although the value of the yield stress seems correct, the value of the fitted plastic viscosity seems to be relatively high compared to more recent studies on the rheological properties of SCC, both theoretically and experimentally. For instance, yield stress of 300 Pa corresponds to a 23 cm slump [21] and Wallevik [66] showed a yield stress range from approximately 10 to 80 Pa and a plastic viscosity range from approximately 20 to 150 Pa.s obtained from measurements in a co-axial viscometer. The above models however showed the first insight into potential casting prediction tools although approximations of the governing equations were applied as well as a mono-dimensional yield criterion.

Since then, further research into prediction of concrete castings has made use of more advanced CFD tools as well as benefitted from increased computational capacity. In the following, an overview of these investigations is presented.

2.6.4.1 SCC Wall Casting

Thrane *et al.* have used FIDAP© to obtain a 3D simulation of the filling process of vertical wall (3 m long, 1.2 m high and 0.30 m wide) [32]. The simulation has been compared to the form filling experiment. The results show good correlation with respect to detection of the free surface location, dead zones and particle paths (*cf.* Fig. 2.12). The simulation approach applied in this study is based on a Galerkin FEM formulation of the Navier-Stokes equation, and a volume-of-fluid representation is used to capture the free surface. A pipe length of 0.40 m was modelled to capture the flow behaviour near the entrance to the form. A constant velocity of 0.35 m/s, was imposed at the start of the pipe corresponding to a filling rate of 10 m^3/h. Instead of a cylindrical inlet, a quadratic inlet was applied to retain the mesh simplicity. A Bingham fluid with yield stress τ_0 = 40 Pa and plastic viscosity μ_p = 20 Pa.s was applied corresponding to slump flow values of 580 mm and $t_{500} \approx 2$ s.

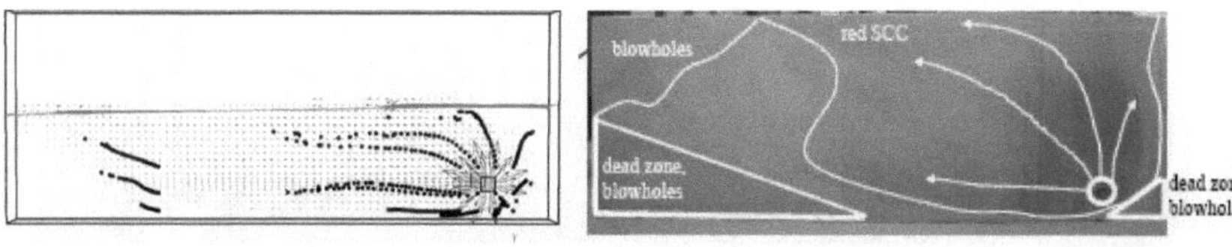

Fig. 2.12 Simulated velocity field and particle paths in the time interval from 120 to 250 s (left). To the right, observed flow behaviour in the experiment (dead zones, particle path of the red SCC, segregation, and surface quality). The injection point is located at the bottom right corner of the formwork (white circle) (from [32]).

Gram [48] also simulated the casting of a wall section in a formwork consisting of film faced plywood panel mounted on steel frames. The formwork was 12.6 m wide, 6.45 m high and 0.27 m thick. It was reinforced with a double mesh made of 12 mm steel bars spaced 200 mm in both directions. The first two modelled lifts hold a volume of V_I = 4.3 m^3 and V_{II} =3.5 m^3. The pump hose placement was horizontally fixed 3 meters from the left end, held right over the surface of the cast concrete. The concrete was pumped in intervals of 3.5 minutes with 10 minutes of resting period during the full hour until the arrival of the next lift. The yield stress of the fresh concrete was determined initially with the LCPC box [25]. Since the pumped concrete was not at rest, effects of thixotropy are not included for the flow model. Initial yield stress levels obtained with the LCPC box were 23.0 Pa for lift I and 21.7 Pa for lift II. After one hour, they were respectively measured at 64.55 Pa and 65.2 Pa. An average slump loss or yield stress increase rate of $\dot{\tau}_0 =$ 0.012 Pa/s was used in the model to simulate structural build-up. Measurements of the casting profile were recorded in three points – left, middle and right end of the formwork every hour before the arrival of the next lift.

The full section was modelled with perfectly smooth walls. A careful placing of the concrete into the formwork was ensured by slow pouring above the concrete surface. The second lift was given different material properties than the first one matching the LCPC-box tests performed at concrete truck arrival time for the new batch. The velocity-dependent plastic viscosity of the concrete has been kept adequately high during the filling process, in order to avoid numerical instabilities created by a 'splash-effect' when the concrete enters the form too quickly. The relatively sparse amount of reinforcement was assumed not to affect the flow of concrete in the formwork and was therefore not included in the model. Fig. 2.13 shows the overall agreement between the actual first casting profiles and the simulations of lift I and II.

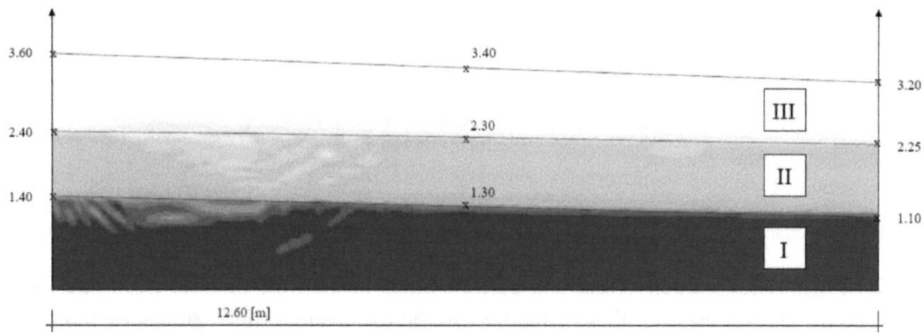

Fig. 2.13 Recorded (I, II, III) and simulated (I and II) casting profiles for the first lifts, x marking the measured depths (from [48])

Van Waarde [46, 47] used the commercial CFD software FLUENT® to perform 2D-simulations of wall castings. The objective of this study was to predict the flow behaviour, to investigate the formwork pressure of SCC as well as to explain local phenomena during casting. The rheological properties of SCC were measured with the BML-viscometer and formwork pressure measurements were carried out at the construction site (four walls). These indicated that hydrostatic values were never reached. In the simulations, the applied casting rate was 1.5 meters/hour and the boundary conditions of the walls were set to 'no-slip'.

The simulations showed areas of high formwork pressure during casting (*cf.* Fig. 2.14) at the points where concrete is poured but no link could be established between simulation results and formwork pressure decrease. The authors concluded that the time-dependent behaviour of SCC should be included in the simulations in order to predict formwork pressure decrease.

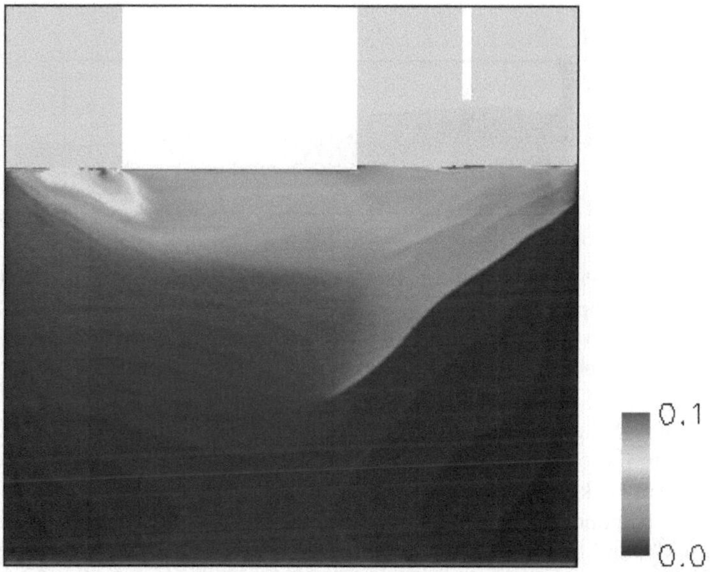

Fig. 2.14 Velocity plot during the casting of a wall (concrete is filled from the top at the right site), from [46]

2.6.4.2 Castings – Consequences of Structural Build Up

Roussel *et al.* [67] have used FLOW3D© to carry out numerical simulations to study the effect of structural build up on the occurrence of distinct layer (cold joint) and formwork pressure. During placing, the fresh SCC behaves as a fluid but, if cast slowly enough or if at rest, it flocculates and builds up an internal structure. A layer of SCC often has a short time to rest and flocculate before a second layer of concrete is cast above it. However, if the fine particles flocculates too much and the apparent yield stress of the concrete increases above a critical value, the two layers do not combine at all and a weak interface is formed, as vibrating is prohibited in the case of SCC (*cf.* 2.15). Loss of resistance of more than 40 % has been reported [67]. According to the delay between the castings of the two layers, it was shown that a distinct interface could or not be spotted.

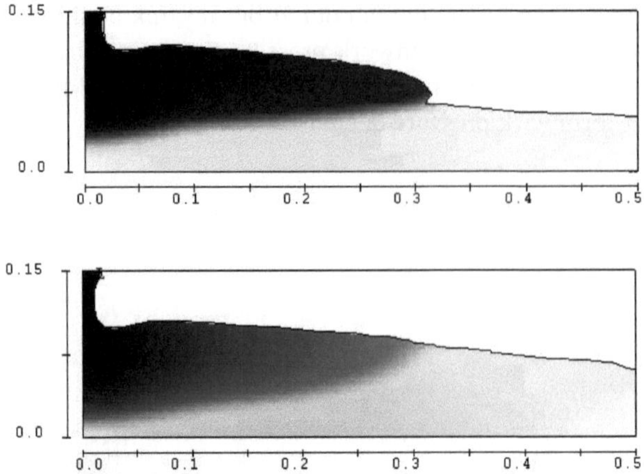

Fig. 2.15 Numerical simulations of the multi-layer casting phenomenon using the thixotropic model proposed in [8] with $\tau_0 = 50$ Pa, $\mu_p = 50$ Pa.s, $A_{thix} = 0.5$ Pa/s, $\alpha = 0.005$. (upper figure) For a 5 min. resting time, the two layers mix perfectly; (lower figure) for a 20 minutes resting time, the two layers do not mix at all (units are meters)

Numerical simulations were also carried out by Ovarlez and Roussel [68] in order to estimate the extent of the zone where SCC is at rest in a formwork according to the formwork geometry and to the casting rate (see Fig. 2.16). The size of the zone where concrete is at rest is indeed of high interest as, in this zone, the ability to withstand the load from concrete cast above it without increasing the lateral stress against the formwork increases over time because of thixotropy [68].

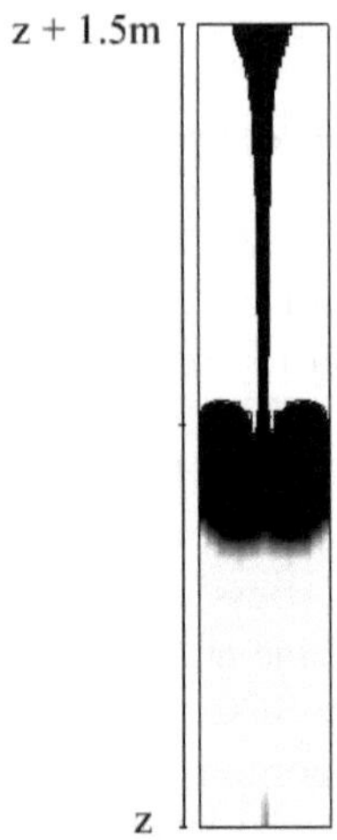

Fig. 2.16 Sheared and resting zones in a wall formwork. SCC is cast from the top of the formwork at a casting rate of 5 m/h. In the black zone, the shear rate is greater than 0.1 s^{-1} (from [68]).

2.6.5 *Industrial Applications*

2.6.5.1 Prediction of Flow in Pre-cambered Composite Beam

Numerical simulations were also applied to an industrial casting of a very high strength concrete (100 Mpa) pre-cambered composite beam by Roussel *et al.* [69]. The results of the simulations carried out for various values of the rheological parameters (Bingham model) helped to determine the value of minimum fluidity needed to cast the element. The mix proportioning of the concrete was done keeping in mind this minimum value and the numerical predictions were finally compared with the experimental observations carried out during two trail castings and the real casting of the two 13 m beams (see Fig. 2.17).

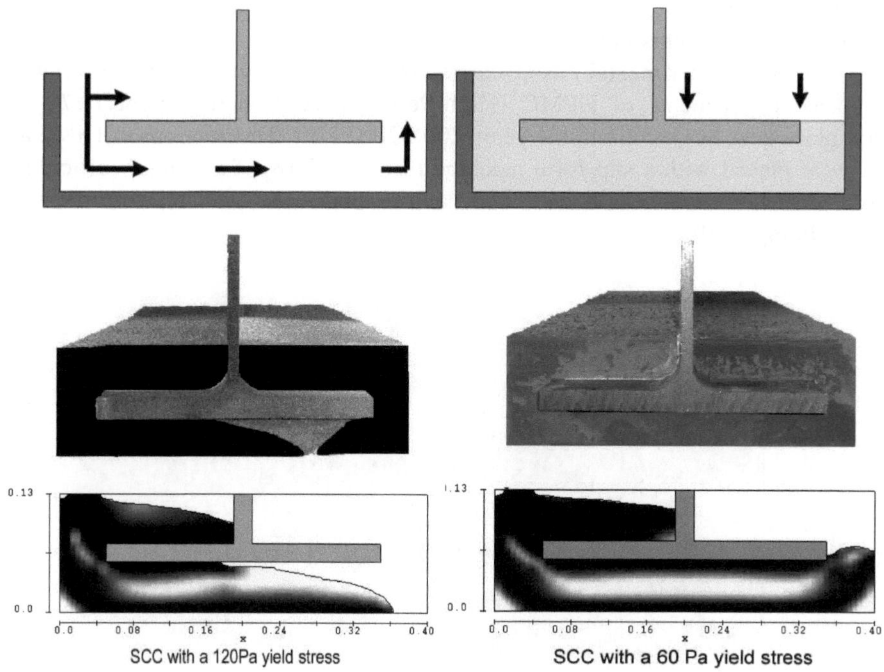

Fig. 2.17 Casting of a concrete pre-cambered composite beam. (upper part left) Schematic showing the first phase of the casting; (upper part right) Schematic showing the second phase of the casting; (lower part) Comparison between experiments and numerical simulations for two SCC with yield stresses equal to 120 Pa (left figures) and 60 Pa (left figures). Black shading emphasizes the casting defect on the upper left picture, Units are meters (from [69]).

Although the assumptions needed to carry out the simulations may be over-simplistic (the rebars and possible thixotropy were not taken into account and only 2D simulations were carried out), a satisfactory agreement was found between the predicted and actual global flow. However, two neglected phenomena locally perturbed the casting: the thixotropic behaviour of the SCC that induced an increase of the apparent yield stress at low casting speed or when the concrete was at rest and the interaction between the largest aggregates and the steel bars that induced some blocking. The fact that only 2D-simulations were carried out was acceptable in the frame of this work since the shape of the element to be cast was suitable for such a simplifying assumption. However, this is in general not the case and the computational time could strongly increase.

2.6.5.2 Flow Simulation of Fresh Concrete under a Slip-Form Machine

In this example of application, the process of casting a thin slab of High-Performance Concrete with a slip-form machine was simulated with a Finite Volume CFD code. Comparison between on-site observations and simulations led to practical improvements of the process.

After a first experimental jobsite carried out in Villeurbanne (2003), a series of full-scale experiments of HPMC (High-Performance Concrete carpet) [70-72] took place near St Quentin in the years 2006-2007 [73]. Test sections of HPC carpet were placed with a slip-form machine (see Fig. 2.18). Two major difficulties were identified: evenness defects and cavities under the steel reinforcement as shown in Fig. 2.19.

Fig. 2.18 Slip-form machine used during the trials performed near A 26 highway, St Quentin, France. Concrete was poured ahead from a truck equipped with a conveyor belt.

Evenness defects were observed when changes occurred in the machine speed (which ranged from 0 to 2 m/min), or when the height of fresh concrete ahead of the machine varied. To solve this problem, a vibration chamber was created at the front of the slip-form. A vertical blade was added in order to impose the concrete height in front of the machine. The extrusion table length was moreover increased from 1.00 m to 1.20 m. However, these changes in the machine had unexpected, severe side effects. After about 60 m of apparently correct placement, defects appeared at concrete demolding, near the right hand side edge of the slab (Fig. 2.19 (right)). Some concrete zones had stiffened in the vibration chamber and were not able to flow properly through the wire mesh.

Fig. 2.19 Casting problems (left) Evenness defects (right) Compaction defects

The computational fluid mechanics finite volume code Flow3D® was used to solve the fluid mechanics equation. 2D-numerical simulations of the flow were carried out (*cf.* Fig. 2.20) with the following assumptions:

11. Flow was considered as sliding at the interface with the extrusion table and sticking at all other interfaces. Hence, the slip-form table is made up with smooth steel, while the ground surface has an uneven profile (*cf.* Chapter 1).
12. The potential thixotropic behaviour of the material was not taken into account. In this work, only a yield stress fluid was considered. To make it even simpler, a Bingham model was used. Thus only two parameters were needed to describe the rheological behaviour: yield stress (1000 Pa) and plastic viscosity (200 Pa.s).
13. The presence of coarse aggregates was ignored. This option could be questioned in this specific case, since the ratio between some critical dimensions, as the clearance between the steel panels and the ground, and the mixture maximum size of aggregate is not very high (20 mm/6 mm = 3.3).

Machine speed and height of concrete ahead of the machine were varied. The simulations and the associated analysis showed that, in order to control the evenness of the slab, it was necessary to provide a regular flow of fresh concrete ahead of the machine. A relationship between height of concrete ahead of the machine and length of the extrusion table preventing the formation of any evenness defects was obtained. It was moreover shown that, with the appropriate table length, it was possible to adapt the machine rate to the concrete supplying rate, with no incidence on the slab evenness.

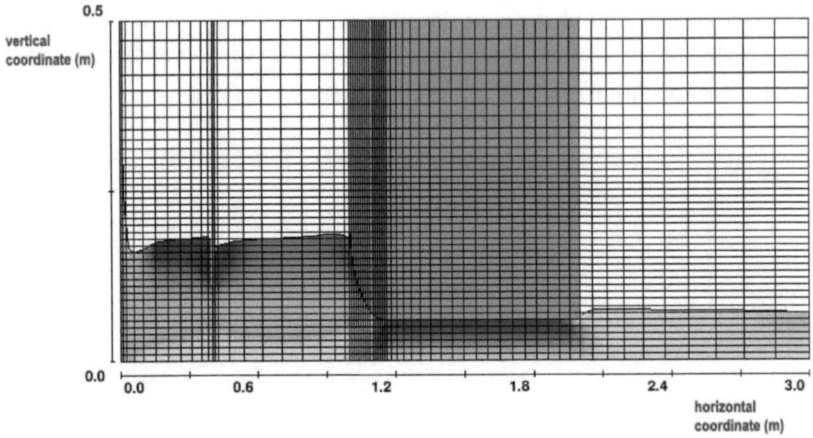

Fig. 2.20 2-D slip-form model. The vertical coordinate is magnified by a factor 3.

Another interesting result came out of the numerical simulations. According to the simulations, when the height of concrete ahead of the machine is high, a rotational motion appears in the vibration chamber (see Fig. 2.21). A non flowing zone may appear in the middle of the vibration chamber above the vibration needles. In such a case, the material stiffens rapidly owing to its thixotropic behaviour [8, 16]. Later in the casting process, the level of concrete ahead of the machine eventually decreases and "blocks" of stiffened concrete are forced to flow under the table. The vibration provided by the needles seems to be insufficient to destructurate the material.

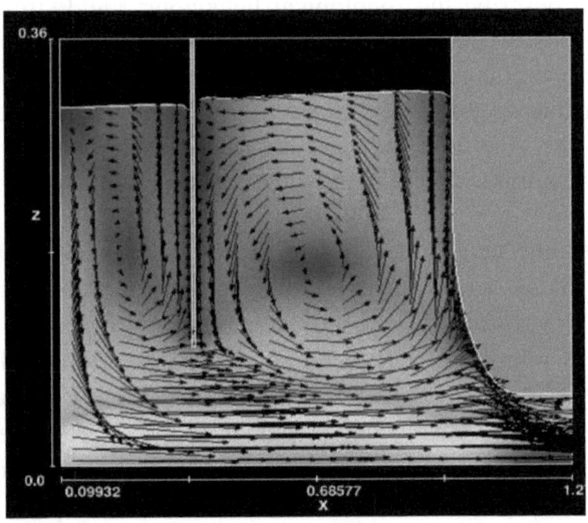

Fig. 2.21 Stream lines in the fresh concrete in the vibration chamber ahead of the machine. A non flowing zone is created in the centre of the chamber.

2.6.5.3 Flow Simulation of Nuclear Waste Disposal Filling

In order to investigate the possibility to store containers with nuclear waste material in horizontal underground tunnels, numerical simulations were used in a project of Delft University of Technology and The Nuclear Research & Consultancy Group (NRG). The space between the tunnel lining and the waste containers could be filled with a cement-based grout (Fig. 2.22). The aim is to create a solid body without voids that can withstand ground settlements.

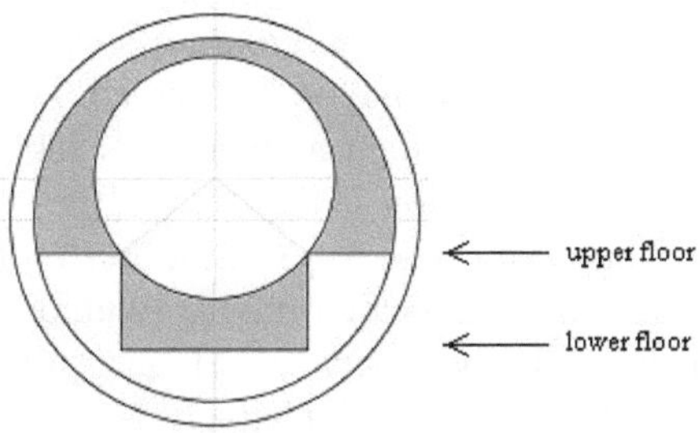

Fig. 2.22 Space of the cross-section of the tunnel to be filled with grout (in grey), from [75]

Since the filling process cannot be visually observed a study was carried out in order to predict the filling by simulating the fresh grout flow with computer software. Bakker [74] combined in his study material testing, downsized tests and the simulation of the grout flow. The objective of the study was to determine a proper filling strategy for this project; the model will be also used to simulate future casting situations. Fig. 2.23 [75] shows a down-scaled Plexiglas section of the gallery (length: 1,92 m; tests were executed in one, two or three section(s)). The Plexiglas model represents the underground tunnel and it was downsized with a factor 12.5. Additional tubes with an outer diameter of 16 cm represent a row of waste containers in the tunnel and are placed inside the Plexiglas setup (inner diameter: 24 cm). This smaller tube rests on line supports along the setup.

Fig. 2.23 Picture of a section of the down-scaledPlexiglas setup (from [75])

Outside the Plexiglas tube, a grid was placed to record the progress of the grout front in time. The progress of the grout front was compared to positions in the simulation. Independent of the length of the setup or the filling method, the inner tube was always immersed in the grout; no air entrapments were observed. At the top of the tube - especially close to the air outlet - a foam-like material appeared. The inner tube was not completely covered (due to the foam). However, since the experiment is a downsized version of the underground tunnel this does not necessarily mean that this also yields for the real situation. The aggregate size distribution (maximum aggregate size of the grout: 4 mm) was determined at different positions along the flow; segregation of the grout (less coarse material was found at the most far end of the injection point) was observed. During the grouting experiment the material is either in constant motion, or is subjected to motion after rest, and finally reaches its destination in the setup. Rheological characteristics (thixotropy, yield value and plastic viscosity) of the fresh grout were determined with the BML-viscometer until 6 hours after mixing (with or without rest periods). Comparable values for the plastic viscosity were found up to a period of 6 hours after mixing (range: 9-13 Pa·s), which is relatively low compared to the plastic viscosity of self-compacting concrete. The yield stress increased during the rest period (3, 4, 5 and 6 hours). When the material was remixed after a resting period, the yield stress was in the range of 0-4 Pa, whereas values between 25 and 47 Pa were found for the material being tested without remixing after a rest period. In the model a constant yield stress and plastic viscosity were assumed as rheological properties of the grout. The thixotropic behaviour was not included in the CFD model.

OpenFOAM, a software package that is based on Computational Fluid Dynamics (CFD), was applied to compare simulations with the flow of the grout in the down-scaled model. The grout was considered a single homogenous fluid with time-dependent characteristics. The Volume Of Fluid method (VOF) was used to track the interface of the two phase flow situation (grout and air). A CFD model was first applied to predict the flow of grout in a funnel test in order to validate the ability of the code to predict the grout flow; a good agreement between the experimental and the simulated funnel times was found. For the down-scaled model, the agreement between the vertical rising of the grout front and the simulations of the gallery was good (*cf.* Fig. 2.24). In the horizontal direction, the grout front in the CFD model was slower than the material in the Plexiglas setup. The differences between the CFD simulation and the flow in the Plexiglas could be caused by the fact that a thin layer of segregating paste preceded the bulk of the grout. This effect was not implemented in the rheological CFD model of the grout.

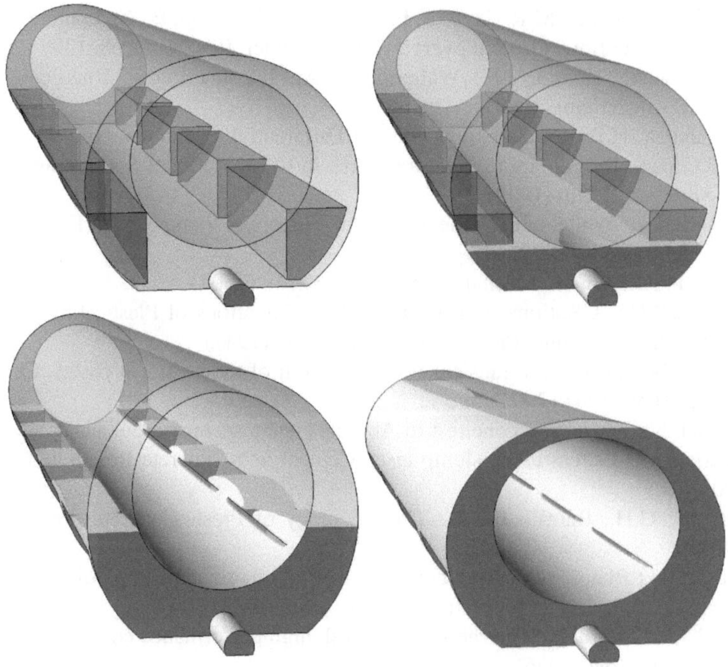

Fig. 2.24 Simulation of grout filling in the down-scaled model [75]

References

[1] Malvern, L.E.: Introduction to the Mechanics of Continuous Medium. Prentice-Hall Inc., New Jersey (1969)

[2] Mase, G.E.: Schaums Outline Series: Theory and Problems of Continuum Mechanics, McGraw–Hill Inc., USA (1970)

[3] Wendt, J.F.: Computational Fluid Dynamics, An Introduction. Springer (1992)

[4] Anderson, J.D.: Computational Fluid Dynamics: The Basics With Applications. Science/Engineering/Math. McGraw-Hill Science (1995) ISBN 0070016852

[5] Hess, J.L., Smith, A.M.O.: Calculation of Potential Flow About Arbitrary Bodies. Progress in Aeronautics Sciences 8, 1–138 (1967), doi:10.1016/0376-0421(67)90003-6

[6] Forrester, T., Johnson, E.N., Tinoco, N., Yu, J.: Thirty years of development and application of CFD at Boeing Commercial Airplanes, Seattle. Computers & Fluids 34, 1115–1151 (2005)

[7] Kotas, P., Tutum, C.C., Šnajdrova, O., Thorborg, J., Hattel, J.: A Casting Yield Optimization Case Study: Forging Ram. International Journal of Metal Casting 4(4) (2010)

[8] Roussel, N.: A thixotropy model for fresh fluid concretes: theory, validation and applications. Cem. Concr. Res. 36, 1797–1806 (2006)

[9] Wallevik, J.E.: Rheological properties of cement paste: thixotropic behavior and structural breakdown. Cem. Concr. Res. 39, 14–29 (2009)

[10] Roussel, N., Geiker, M.R., Dufour, F., Thrane, L.N., Szabo, P.: Computational modeling of concrete flow: General overview. Cem. Concr. Res. 37, 1298–1307 (2007)
[11] Barnes, H.A., Hutton, J.F., Walters, K.: An Introduction to Rheology. Elsevier Science, Amsterdam (1989)
[12] Langtangen, H.P.: Computational Partial Differential Equations, Numerical Methods and Diffpack Programming. Lecture Notes in Computational Science and Engineering. Springer, Berlin (1999)
[13] Tanner, R.I., Walters, K.: Rheology: An Historical Perspective, Elsevier Science B.V., Netherlands (1998)
[14] Irgens, F.: Continuum Mechanics. Springer, Berlin (2008)
[15] Oldroyd, J.G.: A Rational Formulation of the Equations of Plastic Flow for a Bingham Solid. Proc. Camb. Philos. Soc. 43, 100–105 (1947)
[16] Roussel, N.: Steady and transient flow behaviour of fresh cement pastes. Cem. Concr. Res. 35, 1656–1664 (2005)
[17] Roussel, N.: From mini-cone test to Abrams cone test: measurement of cement-based materials yield stress using slump tests. Cement and Concrete Research 35, 817–822 (2005)
[18] Tatersall, G.H., Banfill, P.G.F.: The Rheology of Fresh Concrete. Pitman, London (1983)
[19] Roussel, N.: Rheology of fresh concrete: from measurements to predictions of casting processes. Materials and Structures 40(10), 1001–1012 (2007)
[20] Hu, C., de Larrard, F.: The rheology of fresh high performance concrete. Cem. Concr. Res. 26, 283–294 (1996)
[21] Roussel, N.: Correlation between yield stress and slump: Comparison between numerical simulations and concrete rheometers results. Materials and Structures 39(4), 501–509 (2006)
[22] Roussel, N., Coussot, P.: Fifty-cent rheometer" for yield stress measurements: from slump to spreading flow. Journal of Rheology 49(3), 705–718 (2005)
[23] Thrane, L.N., Szabo, P., Geiker, M., Glavind, M., Stang, H.: Simulation of the test method "L-Box" for Self-Compacting Concrete. Annual Trans. of the Nordic Rheology Society 12, 47–54 (2004)
[24] Nguyen, T.L.H., Roussel, N., Coussot, P.: Correlation between L-box test and rheological parameters of a homogeneous yield stress fluid. Cem. Conc. Res. 36(10), 1789–1796 (2006)
[25] Roussel, N.: The LCPC BOX: a cheap and simple technique for yield stress measurements of SCC. Materials and Structures 40(9), 889–896 (2007)
[26] Poiseuille, J.L.M.: Recherches experimental sur le movement des liquids dans les tubes de très-petits diameters. CR Acad. Sci. Paris 11, 961–967 and 1041–1049 (1840); CR Acad. Sci. Paris 12, 112–115 (1841) (in French)
[27] Macosko, C.W.: Rheology, Principles, Measurements and Applications. Wiley-VCH, New York (1994)
[28] Feys, D., Verhoeven, R., De Schutter, G.: Extension of the Poiseuille Formula for Shear-thickening Materials and Application to Self-compacting Concrete. Appl. Rheol. 18, 62705:1–62705:11 (2008)
[29] Feys, D.: Interactions between rheological properties and pumping of Self-Compacting Concrete, Ph-D Thesis, Ghent University, Ghent (2009)
[30] Kaplan, D.: Pumping of concretes, Ph-D Thesis, LCPC, Paris (2001) (in French)
[31] Chapdelaine, F.: Fundamental and practical study on pumping of concrete, Ph-D Thesis, Université Laval, Laval (2007) (in French)

[32] Thrane, L.N.: Form-filling with Self-Compacting Concrete, Ph-D Thesis, Danish Technological Institute – Technical University of Denmark, Lyngby (2007)
[33] Anderson, J.D.: Computational Fluid Dynamics: The basics with applications, 545 pages. Mcgraw Hills (1995) ISBN 00700116852
[34] Ferziger, J.H., Peric, M.: Computational Methods for Fluid Dynamics, 400 pages. Springer (1999) ISBN, 3540653732
[35] Chung, T.J.: Computational Fluid Dynamics, 2nd edn., 1029 pages. Cambridge University Press (2010) ISBN, 978-0-521-76969-3
[36] Donea, J., Huerta, A., Ponthot, J.-P., Rodrıguez-Ferran, A.: Arbitrary Lagrangian–Eulerian Methods, Encyclopedia of Computational Mechanics. In: Stein, E., de Borst, R., Hughes, T.J.R. (eds.) Fundamentals, vol. 1, John Wiley & Sons, Ltd. (2004) ISBN: 0-470-84699-2
[37] Hirt, C., Nichols, B.: Volume of fluid (vof) method for the dynamics of free boundaries. Journal of Computational Physics 39(1), 201–225 (1981)
[38] Cremonesi, M., Frangi, A., Perego, U.: A lagrangian finite element method for fluid-structure interaction problems with free surface flows. In: Fifth MIT Conference, June 17-19. Massachusetts Institute of Technology, Cambridge (2009)
[39] Mori, H., Tanigawa, Y.: Simulation methods for fluidity of fresh Concrete, Memoirs of the School of Engineering, vol. 44, pp. 71–133. Nagoya University (1992)
[40] Kurokawa, Y., Tanigawa, Y., Mori, H., Nishinosono, Y.: Analytical study on effect of volume fraction of coarse aggregate on Bingham's constants of fresh concrete. Transactions of the Japan Concrete Institute 18, 37–44 (1996)
[41] Christensen, G.: Modelling the flow of fresh concrete: The slump test, Phd thesis, Princeton University (1991)
[42] Thrane, L.N., Szabo, P., Geiker, M., Glavind, M., Stang, H.: Simulation and verification of flow in SCC test methods. In: Proc. of the 4th Int. RILEM Symp. on SCC, Chicago (2005)
[43] Uebachs, S., Brameshuber, W.: Numerical Simulation of the Flow Behaviour of Self-Compacting Concretes Using Fluid Mechanical Methods. In: SCC 2005, Conference Proceedings, The Second North American Conference on the Design and Use of Self-Consolidating Concrete (SCC) and the Fourth International RILEM Symposium on Self-Compacting Concrete, Chicago, Illinois, October 30-November 02, vol. 2. Center for Advanced Cement-Based Materials, Evanston (2005)
[44] Uebachs, S., Brameshuber, W.: Investigations of the Fluid Structure Interac-tion on the Flow Behaviour of Self Compacting Concrete Regarding the Numerical Flow Simulation. In: Schutter de, G., Boel, V. (eds.) 5th International RILEM Symposium on Self-Compacting Concrete, SCC 2007, Ghent, Belgium, September 3-5, vol. 1, pp. S.429–S.435. RILEM Publications, Bagneux (2007)
[45] Tregger, N., Ferrara, L., Shah, S.P.: Identifying cement rheological properties from the mini-slump-flow test. ACI Materials Journal 105(6), 558–566 (2008)
[46] Waarde, F.V.:: Formwork pressures when casting self compacting concrete, Master thesis, Delft University of Technology, Department of Structural and Building Engineering (2007)
[47] Waarde, F., Koenders, E.A.B., Lycklamaà Nijeholt, J.A., Walraven, J.C.: Theoretical and practical investigations on SCC formwork. In: RILEM-Conference SCC, Ghent (2000, 2007)
[48] Gram, A.: Numerical modelling of self-compacting concrete flow – discrete and contiunous approach, lic. thesis Royal Institute of Technology, Sweden (2009)

[49] Bras, A.: Grout optimization for masonry consolidation, Dissertation, Universidade Nova de Lisboa, Portugal (waiting for dissertation discussion) (2010)

[50] Nguyen, V.H., Rémond, S., Gallias, J.L., Bigas, J.P., Muller, P.: Flow of Herschel–Bulkley fluids through the Marsh cone. Journal of Non-Newtonian Fluid Mechanics 139(1-2), 128–134 (2006)

[51] Le Roy, R., Roussel, N.: The Marsh Cone as a viscosimeter: Theo-retical Analysis and Practical Limits. RILEM Materials and Structure 37, 25–30 (2004)

[52] Roussel, N., Le Roy, R.: The Marsh cone: a test or a rheological apparatus? Cement and Concrete Research 35(5), 823–830 (2005)

[53] Hu, C.: Rheologie des betons fluides, Etudes et Recherches des Laboratoires des Ponts et Chaussees, OA 16, Paris, France, 203 p. (1995)

[54] Wallevik, J.E.: Development of Parallel Plate-Based Measuring Sys-tem for the ConTec Viscometer. In: Proceedings of the 3rd International RILEM Symposium on Rheology of Cement Suspensions such as Fresh Concrete, August 19-21. RILEM Publications S.A.R.L., Reykjavik (2009) ISBN: 978-2-35158-091-2

[55] `http://www.vvpf.net`

[56] Rivera, C.A., Heniche, M., Takenaka, K., Tanguy, P.A.: Finite element modeling of the laminar and transition flow of the Superblend dual shaft coaxial mixer on parallel computers. Chemical Engineering Science 64, 4442–4456 (2009)

[57] Chevoir, F.: Rhéologie des matériaux granulaires, Congrès Matériaux, Nantes (France), Octobre 18-22, pp. 18–22 (2010)

[58] Rui, H., Tabata, M.: A second order characteristic finite element scheme for convection-diffusion problems. Numer. Math. 92, 161–177 (2002)

[59] Xiaoxia, D., Xiaoliang, C.: The iterative penalty method for Stokes equations using Q1-P0 element. Applied Mathematics and Computation 201, 805–810 (2008)

[60] Girault, V., Glowinski, R., Lopez, H., Vila, J.P.: A boundary multiplier/fictitious domain method for the steady incompressible Navier-Stokes equation. Numer. Math. 88, 75–103 (2001)

[61] Chang, K.S., Kwak, D.Y.: Discontinuous bubble scheme for elliptic problems with jumps in the solution. Comput. Methods Appl. Mech. Engrg. 200, 494–508 (2011)

[62] Berger, M., Oliger, J.: Adaptive mesh refinement for hyperbolic partial differential equations. J. Comput. Phys. 53, 484–512 (1984)

[63] Dean, E.J., Glowinski, R., Guidoboni, G.: On the numerical simulation of Bingham visco-plastic flow: Old and new results. J. Non-newtonian Fluid Mech. 142, 36–62 (2007)

[64] Weber, R.: The Transport of Concrete by Pipline, Translated by C. van Amerongen in Cement and Concrete Association Translation 129 (1968)

[65] Kitaoji, H., Tanigawa, Y., Mori, H., Kurokawa, Y., Urano, S.: Flow simulation of fresh concrete cast into wall structure by viscoplastic divided space element method. Transactions of the Japan Concrete Institute 16, 45–52 (1996)

[66] Wallevik, O.: Rheology – a scientific approach to develop self-compacting concrete. In: Proceedings of the 3rd International RILEM Symposium on Self-Compacting Concrete, RILEM PRO 33, Reykjavik, Iceland, pp. 23–31 (2003)

[67] Roussel, N., Cussigh, F.: Distinct-layer casting of scc: the mechanical conse-quences of thixotropy. Cement and Concrete Research 38, 624–632 (2008)

[68] Ovarlez, G., Roussel, N.: A physical model for the prediction of lateral stress exerted by self-compacting concrete on formwork. RILEM Materials and Structures 39(2), 269–279 (2006)

[69] Roussel, N., Staquet, S., D'aloia Schwarzentruber, L., Le Roy, R., Toutlemonde, F.: SCC casting prediction for the realization of prototype VHPC-precambered composite beams. Materials and Structures 40(9), 877–887 (2007)

[70] de Larrard, F.: La moquette en BHP - Un nouveau concept de couche de roulement. Note Technique, Bulletin des Laboratoires des Ponts et Chaussées (223), 105–110 (1999) (in French)

[71] de Larrard, F.: High-Performance Concrete Carpet: a hydraulic flexible wearing course. Part I: design. Scientific Note, International Journal of Road Materials and Pavement Design 6(4), 533–548 (2005)

[72] de Larrard, F., Kerzreho, J.-P., Potier, J.-M., Baroin, L., Abdo, J.: High-Performance Concrete Carpet: a hydraulic flexible wearing course. Part II: full-scale testing. Scientific Note, International Journal of Road Materials and Pavement Design 6(4), 549–564 (2005)

[73] de Larrard, F.: Moquette BHP - Expérimentations de Saint Quentin 2006-2007 - Autoroute A 26 SANEF - Rapport de synthèse. Convention LCPC/SANEF, 46 p. (Décembre 2007) (in French)

[74] Bakker, P.: Grout backfill process of a nuclear waste disposal gallery - A study into the most suitable filling strategy, Master thesis, Delft Univer-sity of Technology, Department of Structural and Building Engineering (2009)

[75] Bakker, P., Ramohalli Gopala, V., Lycklamaà Nijeholt, J.A., Koenders, E., Grünewald, S., Walraven, J.: Simulation and Testing of the Grout Backfill Process in a Case-Study Related to a Nuclear Waste Disposal Gallery. In: Design, Production and Placement of Self-Consolidating Concrete. Rilem Bookseries, vol. 1, pp. 197–208 (2010) ISBN: 978-90-481-9663-0

[49] [illegible] SPC [illegible] [illegible] Structure [illegible]

[50] [illegible] Nav[illegible] [illegible]

[illegible]

Chapter 3
Simulation of Fresh Concrete Flow Using Discrete Element Method (DEM)

Viktor Mechtcherine[1,*], Annika Gram[2], Knut Krenzer[3], Jörg-Henry Schwabe[4], Claudia Bellmann[1], and Sergiy Shyshko[1]

[1] Institute of Construction Materials, Technische Universität Dresden, Germany
[2] Swedish Cement and Concrete Research Institute CBI, Sweden
[3] Institut für Angewandte Bauforschung Weimar gGmbH
(formerly Institut für Fertigteiltechnik und Fertigbau Weimar e.V.), Germany
[4] University of Applied Science, Jena, Germany

3.1 Introduction

The behaviour of fresh concrete during its mixing, transport, placement, and compaction can ultimately have significant effects on its mechanical performance, durability, surface appearance, and on its other properties after hardening. In concrete construction many problems result from the improper filling of formwork, insufficient de-airing, concrete segregation, *etc.* The importance of these issues has increased year after year since formwork is becoming continually more complex. Steel reinforcement has become denser, and the range of workability has been considerably broadened by the use of self-compacting concrete (SCC) and other novel concrete materials. Consequently, on the one hand, modern material design must match particular demands resulting from the geometrical and technological conditions to which the material is subjected. On the other hand, the concrete working techniques and, in some cases, the geometry of structures can be optimised in considering the usage of particular concretes with their special rheological properties. So, in order to build concrete structures efficiently and with high quality, the consistency of the fresh concrete should comply with the requirements posed by the structure's geometry and by the methods of transport, placing, and compaction. Computer simulation of fresh concrete behaviour and the working processes could provide a powerful tool in optimising concrete construction and developing new concrete technologies [1].

As a complement to laboratory experimentation, discrete numerical simulation applied to granular materials provides insight into the meso-structure on the same scale as the grains and improves our understanding of the origin of macroscopic material behaviour. A description of various discrete simulation methods used in the mechanics literature can be found in [2]. This chapter focuses on the application of Discrete Element Method (DEM) in simulating the flow of fresh

* Corresponding author.

N. Roussel and A. Gram (eds.), *Simulation of Fresh Concrete Flow*,
RILEM State-of-the-Art Reports 15,
DOI: 10.1007/978-94-017-8884-7_3, © RILEM 2014

concrete. In particular a so-called particle model approach, which is a variation of the DEM, is considered. This approach uses simple basic elements – spheres and walls, which makes the computation relatively simple and fast. The great advantage of DEM is that it provides an opportunity to display discreetly the movement of the concrete mixture as a whole, and of its individual components.

The concrete mixture is formed by a large number of particles connected among each other and to the model walls in accordance with laws of the defined contact behaviour. Thus, concrete technology's relevant processes and phenomena, such as mixing, compaction, de-airing, sedimentation, fibre distribution, orientation, *etc.* could be analyzed and taken into account in solving various problems.

Based on work by Chu *et al.* [3] and Chu and Machida [17], a 3D DEM using a 3D particle flow code program, PCD^{3D}, was applied in a preliminary study by Noor and Uomoto [4] to simulate the flow of SCC during various standard tests: the Slump Flow, L-box and V-funnel procedures. As opposed to an approach whose basis is a continuum, DEM was selected and was observed by the authors to reproduce the behaviour of fresh concrete in a qualitatively correct manner. As a compromise between modelling aggregate movement and limiting computation time, the material was divided into mortar and coarse aggregates larger than 7.5 mm. The method, proposed by Noor and Uomoto, was also adopted by Petersson and Hakami [5] and Petersson [6] in simulating SCC flow during L-box and slump flow testing, and J-ring and L-box testing, respectively. They found 3D- and, depending on the type of problem, 2D-simulations to be appropriate.

More recently it was shown by Mechtcherine and Shyshko [7-9] that this numerical technique allows the simulation of the behaviour of fresh concrete with varied consistencies during transport, placement, and compaction. Processes such as casting, compaction of ordinary concrete, wet spraying and extrusion have been simulated as well. In the case of fibre-reinforced materials the effects of concrete consistency and the working process on the orientation of fibres have been of particular interest. The correlation between mix design and rheology was also investigated through the effect of adding large aggregates or fibre. Furthermore, first attempts towards modelling air inclusions and de-airing were carried out.

DEM was also applied by Schwabe *et al.* [10, 11] in modelling and analysing the blending of the grain ingredients within a concrete mixer. The mixing container and the mixing tools consist of PFC-walls. The movements of the tools are assigned to these walls so that the walls interact with the particles. The different grain fractions are filled successively into the mixer and the analysis of the mixture quality is attained by a virtual extraction of a sample from within the mixing box. The numerical results obtained were validated by the use of an experimental environment in which mixture quality and material flow were measured. Experimental analyses with different construction designs in planetary mixers at real scale confirmed the quality of the simulations.

The purpose of this chapter is to summarise the state-of-the-art in the field of simulating concrete flow using DEM. Particular focus is directed at the presentation of the mathematical methodology and the rheological modelling, including the identification of the model parameters for the simulation of fresh concrete as an important step toward the reliable quantitative analysis of the working processes of fresh concrete using the Discrete Element Method. Furthermore, some representative and promising examples of industrial applications of this relatively new approach are shown.

3.2 Discrete Element Method

3.2.1 Governing Equations

The Particle Method referred to in this chapter is a variation of the Discrete Element Method and enables the modelling of the movement (translation and rotation) of distinct particles, including their interactions as well as their separation and automatic contact detection [12]. This method was originally developed as a tool to perform research into the behaviour of granular material. A fundamental assumption of this method is that the material consists of separate discrete particles. Forces acting on each individual particle are computed according to the relevant laws of physics.

The particles themselves are defined to be rigid. Their interaction is treated as a dynamic process with a developing state of equilibrium whenever the internal forces are in balance. The contacts between neighbouring particles occur only at one point at a given time. The calculations alternate between the application of Newton's Second Law with respect to the motion of particles and the force-displacement law at the contacts [18].

Each individual particle moves according to Newton's Second Law and the equation for torque. Thus the displacements and rotations of the particles are calculated according to the following governing Equations (1) and (2).

$$F_i = m \cdot (\ddot{x}_i - g_i) \tag{1}$$

$$M_i = \mathrm{I} \cdot \dot{\omega}_i \tag{2}$$

where F_i denotes the contact force vector, m is the mass of the particle, $\ddot{x}_i$ is the translational acceleration, and g_i is the acceleration of gravity. The torque M_i is the resultant moment acting on the particle, comprised of the moment of inertia I and the angular acceleration of the particle $\dot{\omega}_i$. The translational motion of the centre of mass of each particle is described in terms of position x_i, velocity $\dot{x}_i$, and acceleration $\ddot{x}_i$; the rotational motion of each particle is described in terms of its angular velocity ω_i and its angular acceleration $\dot{\omega}_i$, see, e.g., [15].

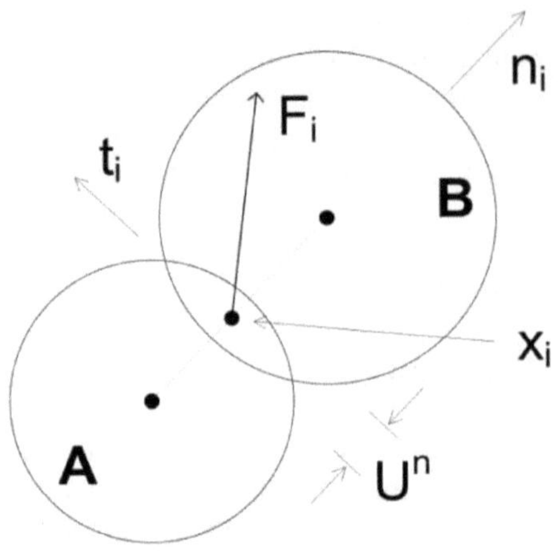

Fig. 3.1 Contact model of Particle Flow Code according to [15]

The contact force vector F_i, representing the action of particle A on particle B in particle-to-particle contact is shown in Fig. 3.1 and also represents the action of the wall on the particle for wall-to-particle contact. It can be resolved into its normal and shear components with respect to the contact plane according to Equation 3:

$$F_i = F_i^n + F_i^s \tag{3}$$

where F_i^n and F_i^s denote the normal and shear component vectors, respectively.

The force-displacement law relates these two force components to the corresponding components of the relative displacement.

3.2.2 Solution Procedure

As described in Section 3.2.1 Newton's law (*cf.* Eq. (1)) is used to determine the motion of each particle arising from the contact and the forces acting upon it as a body. The dynamic behaviour is represented numerically by a time-stepping algorithm which assumes that the velocities and accelerations are constant within each time step.

The equations of motion are integrated using a centred finite difference procedure [15]. Velocities and angles are calculated halfway through the time step at $t \pm \Delta t/2$, Δt being the size of the step. Displacements, accelerations, angular velocities, forces and moments are computed at the primary intervals of $t \pm \Delta t$. The accelerations are calculated as

$$\ddot{x}_i^{(t)} = \frac{1}{\Delta t} \cdot \left(\dot{x}_i^{(t+\Delta t/2)} - \dot{x}_i^{(t-\Delta t/2)} \right) \tag{4}$$

$$\ddot{\omega}_i^{(t)} = \frac{1}{\Delta t} \cdot \left(\dot{\omega}_i^{(t+\Delta t/2)} - \dot{\omega}_i^{(t-\Delta t/2)} \right) \tag{5}$$

Inserting the expressions above into the governing equations for particle displacements (Eq. (1)) and rotation (Eq. (2)) we get:

$$\dot{x}_i^{(t+\Delta t/2)} = \dot{x}_i^{(t-\Delta t/2)} + \left(\frac{F_i^{(t)}}{m} + g_i \right) \cdot \Delta t \tag{6}$$

$$\omega_i^{(t+\Delta t/2)} = \omega_i^{(t-\Delta t/2)} + \left(\frac{M_i^{(t)}}{\mathrm{I}} + g_i \right) \cdot \Delta t \tag{7}$$

Finally, the positions are updated according to:

$$x_i^{(t+\Delta t)} = x_i^{(t)} + \dot{x}_i^{(t+\Delta t/2)} \cdot \Delta t \tag{8}$$

The force-displacement law (*cf.* Chapter 3.3.3.) is then used to update the contact forces arising from the relative motion at each contact. This process is based on the relative motion between the two entities in contact and the constitutive model used in the particular contact. Next, the law of motion is again applied to each particle to re-determine its velocity and position. This is based on the resultant force and moment arising from the contact forces and any other force acting on the body, *e.g.*, gravity.

3.2.3 Software Used in Concrete Engineering

3.2.3.1 Particle Flow Code (PFC) from ITASCA

ITASCA was founded 1981 by members of the University of Minnesota, USA to provide services in rock mechanics, numerical modelling of geotechnical environments, and underground space use. In 1994 Dr. Peter Cundall and his team of researchers developed PFC^{2D} and PFC^{3D} (shown in Fig. 3.2), the first industrially usable DEM simulation tool [18]. The origins of rock and soil mechanics were extended with applications like bulk movement and fracture mechanics.

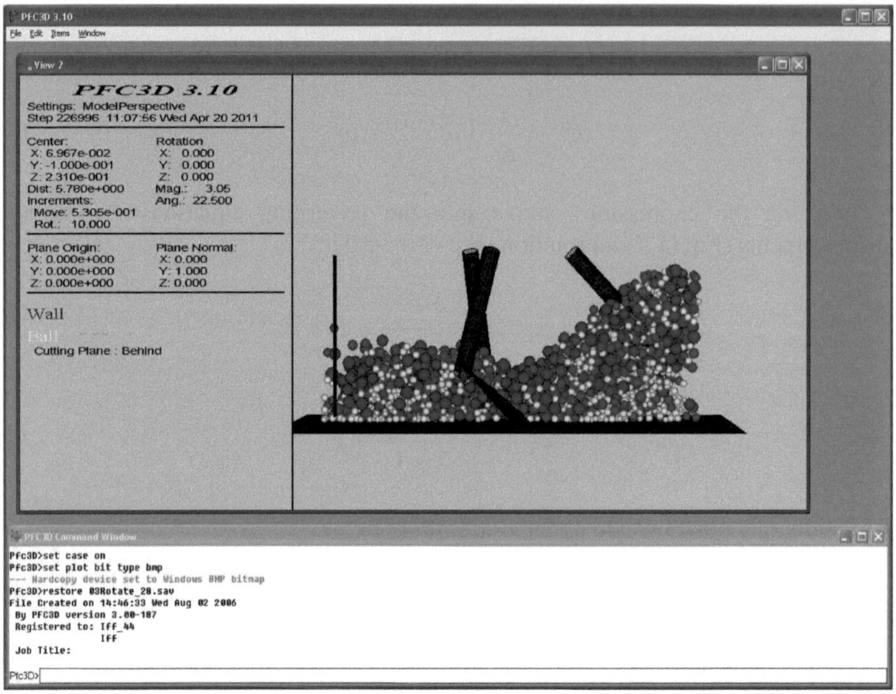

Fig. 3.2 Graphic user interface of PFC^{3D} 3.1, showing the simulation of a lab mixer (figure courtesy of IFF Weimar e.V.)

The basis of the program is an elementary programming language to model the simulation elements, manage the computations, and process the analyses. This simplicity enables direct access to all the information in the simulation. The basic elements are spheres, which can be combined to form complex structures, or "clumps". The essential part of enhanced contact laws can be introduced by user defined models written in C++. Also there are basic approaches for the coupling of particle movement (DEM) with fluid flow (CFD). A parallel computation on different CPU cores can be performed by dividing the model into several parts and solve the sub-domain on each CPU.

The first applications of DEM on fresh concrete were all processed with PFC: 1996 in Japan [3], 2001 at the CBI Stockholm [5], 2005 at the IFF Weimar e.V. [11] and 2006 at the TU Dresden [7].

3.2.3.2 EDEM from DEM Solutions

DEM Solutions Ltd. was founded 2002 in Edinburgh. Since 2005 EDEM has been offered as the software for the commercial application of DEM. The main purpose of the software is the modelling of bulk material flow.

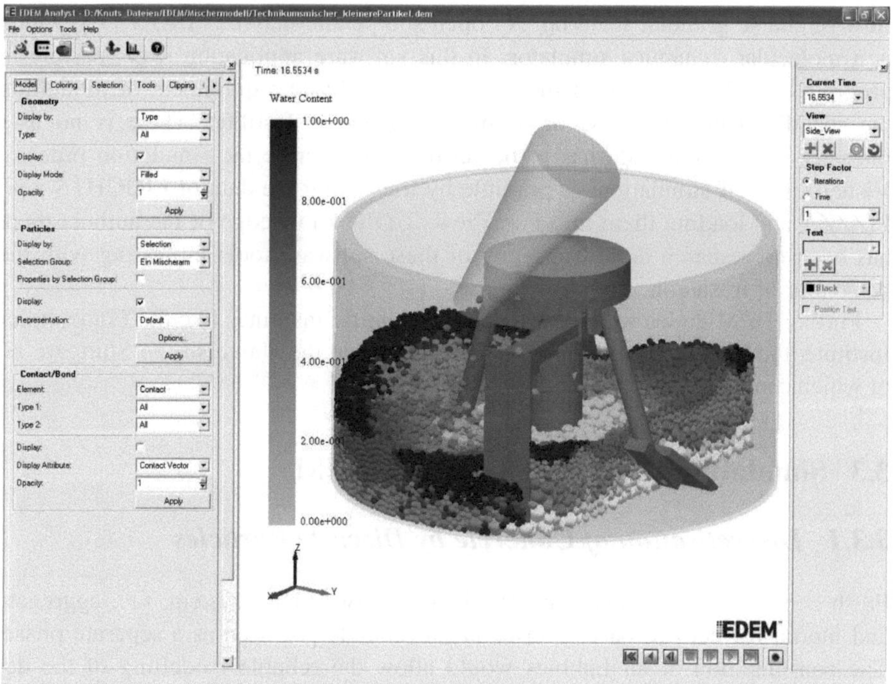

Fig. 3.3 Graphic user interface of EDEM, showing the simulation of a lab mixer (picture courtesy of IFF Weimar e.V.)

The software offers a modern graphic user interface (cf. Fig. 3.3) to process the complete simulation procedure from model generation to simulation and extensive analyses. The philosophy of complex particle shapes with clumps is implemented directly, including particle shape import and export. A big advantage is the capability to import CAD geometry data and therefore have the possibility of easy inclusion of complex machine designs. It is also possible to couple EDEM with the CFD software Fluent to model complex processes influenced by particles and fluid flow. EDEM can be used on multi-core CPUs without an additional model split up by the user. The application to fresh concrete is carried out at the IFF Weimar e.V. and at the FH Jena, among others.

3.2.3.3 Alternative DEM Software

Besides the software distributions presented above, there are some further DEM software tools. Commercial programmes available are, for example, ELFEN and Chute Maven, both of which find their major applications in the bulk material flow of coarse granular material. An open source alternative is LIGGHTS, which is a molecular dynamics simulator. In this software application it is possible to program user-defined contact models so that the basic requirement of modelling the specific material behaviour of fresh concrete is fulfilled. There is no user interface and all the code has to be scripted to describe the simulation process. Visualization of simulations is possible by translating the data of LIGGHTS with PIZZZA and loading them into ParaView. To the knowledge of the authors there has been no research done using any of these software tools in dealing with the behaviour of fresh concrete.

Further DEM codes were developed in scientific institutes like the Frauenhofer Institute, the Research lab for concrete in Japan and the University of Stuttgart, all of which cannot be reviewed in this chapter.

3.3 Simulating Concrete Flow Using DEM

3.3.1 Discretisation of Concrete by Discrete Particles

Fresh concrete is generally considered to be a two-phase system, *i.e.*, aggregate and mortar (or cement paste); however, air can also be seen as a separate phase. The consideration of air bubbles would allow the reliable modelling of the de-airing behaviour of concrete. In any case each of these phases is simulated at its lowest level by circular, spherical or clumped particles with specific properties according to the modelled phase. The interaction between individual particles is controlled by appropriate constitutive relations. In contrast to a more general DEM, only two basic elements are used in the Particle Method: circular (2D) or spherical (3D) particles and walls. Circular or spherical particles, depending on the number of dimensions chosen, are used to render the concrete meso-structure discrete, *i.e.,* as coarse aggregates and fine mortar, where walls are used to simulate the boundaries. The use of simply shaped basic elements renders in turn the contact detection simple and the calculation fast. If more complex geometries are to be simulated – like fibres or non-spherical aggregates, a number of particles can be rigidly interconnected providing necessary geometries (*cf.* Fig. 3.4). The group at the TU Dresden has already successfully modelled both simple tests (slump flow) and the filling of a rectangular mould with SCC containing fibres [8]. The grain sizes, the content of the aggregate, and the grain or fibre distribution in the mixture at the beginning of the simulation have been generated for the concrete volume under consideration in the pre-processing by a corresponding subroutine. Depending on the material under consideration, an appropriate specific mass is assigned to the particles.

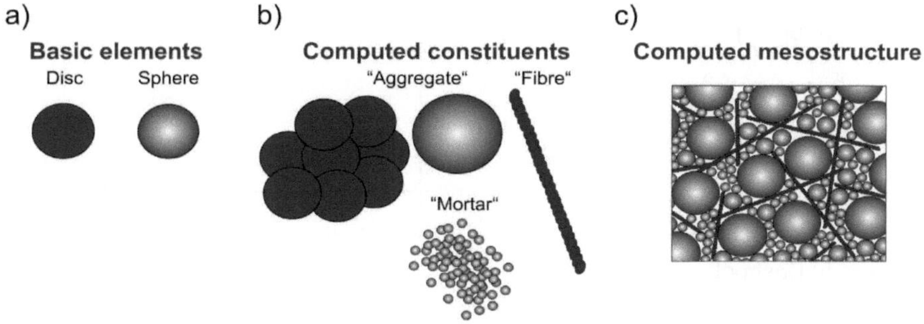

Fig. 3.4 Schematic view of the (a) basic elements, (b) computed constituents of concrete, and (c) discrete rendering of fibre reinforced concrete (figure courtesy of TU Dresden)

3.3.2 Rheological Model

As stated by Malkin [19], rheology is the theory studying the properties of matter in determining its reaction to deformations and flow. Structural changes in materials under the influence of applied forces result in deformations which can be modelled as superimpositions of viscous, elastic and plastic effects. It is useful to introduce basic rheological models as a basis for description of complex material behaviour. These basic models are uni-dimensional models describing elements of rheological behaviour mathematically. They may consist of the following analogies describing a material, here as mechanical models shown in Fig. 3.5 [16].

External forces acting upon a material may result in deformation that can be either elastic, as in the case of a spring (deformation completely recoverable when the force is released) or plastic, as given by the slider (deformation irrecoverable), or viscous (rate-dependent). Viscosity may be visualized as a dashpot, the stress being proportional to the shear rate. In the case of a Newtonian fluid and one dimension, this can be written as

$$\tau = \eta \cdot \dot{\gamma} \tag{9}$$

with τ being the shear stress, η the viscosity and $\dot{\gamma}$ the shear rate. The stress to shear rate ratio, the slope of the function (*cf.* Fig. 3.5b), is the viscosity.

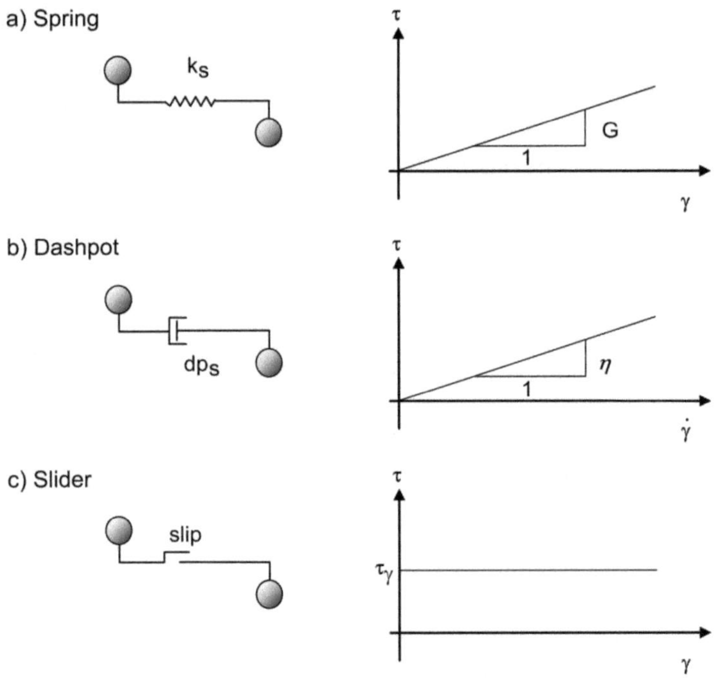

Fig. 3.5 Diagrammatic representation of basic rheological models, describing (a) elastic, (b) viscous and (c) plastic behaviour [16]

Concrete and other concentrated suspensions are most often modelled as so-called Bingham materials. These are materials showing little or no deformation up to a certain level of stress. Above the yield stress τ_0 the material flows. These materials are called viscoplastic or Bingham plastics after E.C. Bingham, who was the first to use this description on paint in 1916, Macosko [20]. With G being the spring constant and τ, γ, and $\dot{\gamma}$ being the shear stress, shear deformation and shear rate of the material, respectively, eqs. (10) and (11) can be written:

$$\tau = G \cdot \gamma \qquad \text{for } \tau < \tau_0 \tag{10}$$

$$\tau = \tau_0 + \mu \cdot \dot{\gamma} \qquad \text{for } \tau \geq \tau_0 \tag{11}$$

where μ is the plastic viscosity.

The yield stress defines the deformability of the concrete, which is one parameter describing workability. Visualized as well by Roussel [21], the shearing behaviour of a Bingham material can be represented by a dashpot, a spring, and a slip function (*cf.* Fig. 3.6).

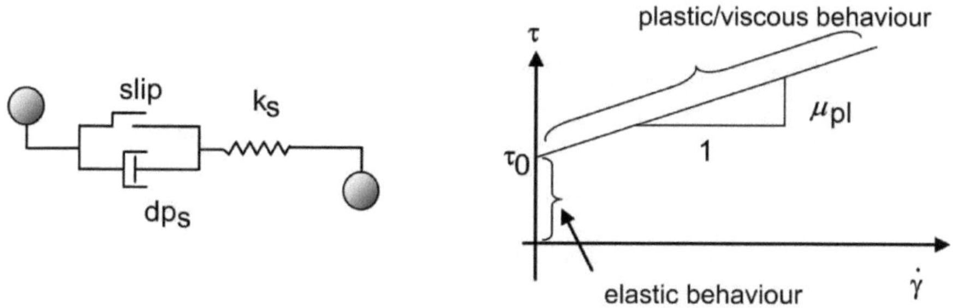

Fig. 3.6 A Bingham material may be described in terms of a spring, a dashpot and slip function [14]

For numerical reasons, the simulated spring is very stiff, for the theoretical model, it is infinitely stiff. The threshold value of the slip function is at the level of the yield stress. Once it is attained, the material will move according to the plastic viscosity of the dashpot (corresponding to the slope of the function). The stress-to-shear-rate ratio is called the apparent viscosity and is equivalent to the viscosity η as defined by Newton's law. The apparent viscosity is always higher than the plastic viscosity μ, but it approaches the plastic viscosity for very high shear rates. The stopping criterion of the flow for a Bingham liquid is the yield stress.

3.3.3 Constitutive Relationships

Constitutive relationships associated with the Bingham formula were developed by the group at the TU Dresden and implemented into the Particle Flow Code in order to describe the interaction between two neighbouring particles in simulating fresh concrete. The corresponding rheological models for the normal and tangential direction are shown in Fig. 3.7. They consist of the basic rheological elements spring, dashpot, and slider, which respectively represent the elastic, viscous and frictional components of the particles' interaction (*cf.* Section 3.3.2).

However, the interaction model includes also the element "contact", positioned serially in line with the basic rheological elements. This additional element enables the definition of the strength of the contact, the simulation of the loss of an old interaction by reaching a certain distance between two particles, and the formation of a new interaction. Fig. 3.8 shows schematically two types of force-displacement relations as introduced by Shyshko and Mechtcherine [13] for the contact elements in the normal direction and subsequently used in numerical investigations.

The first version (CM1) of the force-displacement law in the normal direction includes two main modes: compression and tension (*cf.* Fig. 3.8). The compression mode is defined by a fixed value of stiffness, while the force-displacement curve in tension mode linearly ascends to a defined ultimate force (the bond strength) and then linearly decreases down to zero in a kind of softening regime. When the tensile force becomes zero, the particles lose contact.

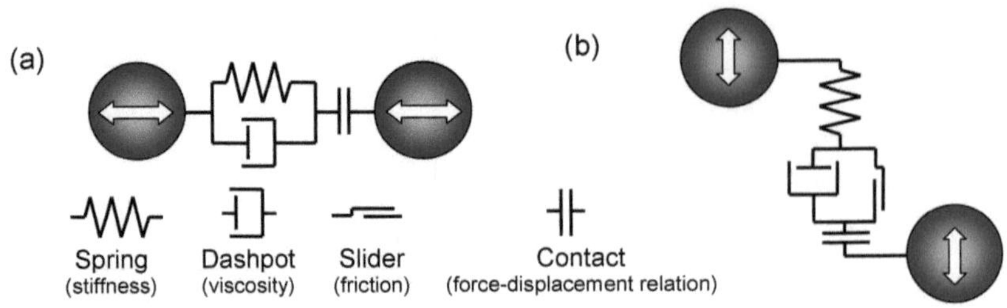

Fig. 3.7 Model for particle interaction: (a) normal direction and (b) tangential direction [13]

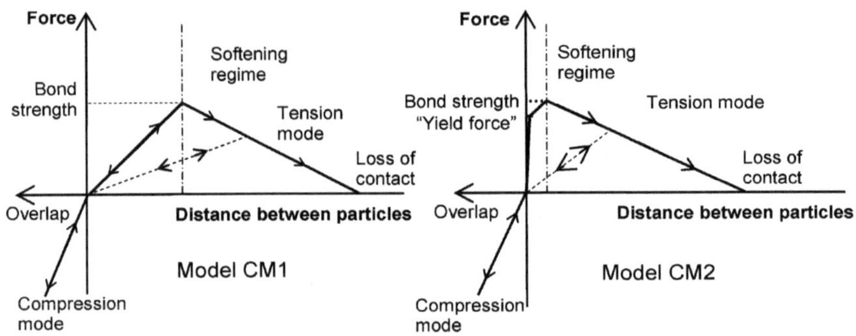

Fig. 3.8 Two types of force displacement relation for contact elements (normal direction) [13]

Since model CM1 for particle interaction in the normal direction does not include frictional elements (*cf.* Fig. 3.7a), the modelling of the characteristic behaviour of fresh concrete related to the yield stress τ_0 might not be accurate enough if the force-displacement relation as described in the previous paragraph is used. Therefore, a slightly modified contact model CM2 was proposed as an alternative. The contact between neighbouring particles in tension is defined at small deformations by a very steeply ascending branch, *i.e.*, there is practically no deformation until a given force value (here "yield force") is reached. After reaching this force level there is only a slight increase in tensile force while the corresponding deformations increase rapidly. The descending branch of the force-displacement relation does not defer from that of the model CM1.

3.3.4 Parameter Estimation

A major challenge of particle simulation is the calibration of the material model. The parameters of material composition such as content of fines, water, and additives can only be considered by an appropriate selection of the general model parameters describing the interactions between individual particles. The calibration of parameters for the contact models is usually performed with simple reference experiments that are processed in the lab and modelled in simulation. The material parameters in the simulation of the reference experiment are iteratively adjusted until the results of the real experiment and the simulation match.

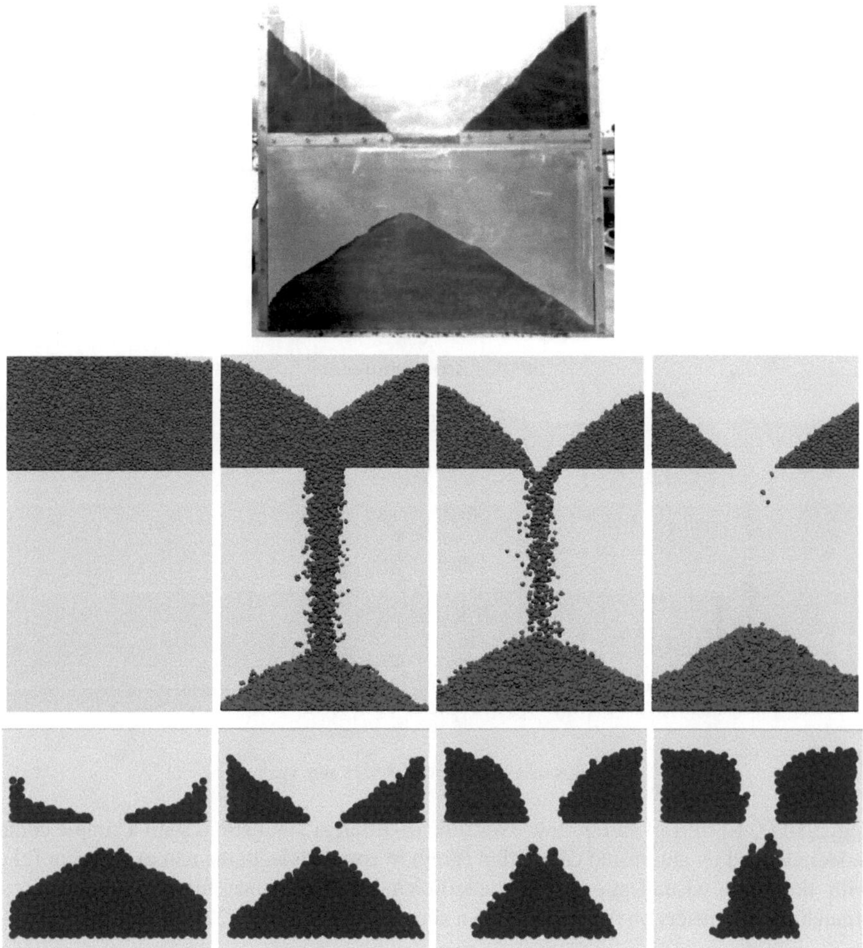

Fig. 3.9 (a) Real experiment for angle of repose and tearoff edge; (b) process stages of one simulation; (c) simulation results for various sets of model parameters (rolling resistance and friction); picture courtesy of IFF Weimar e.V.

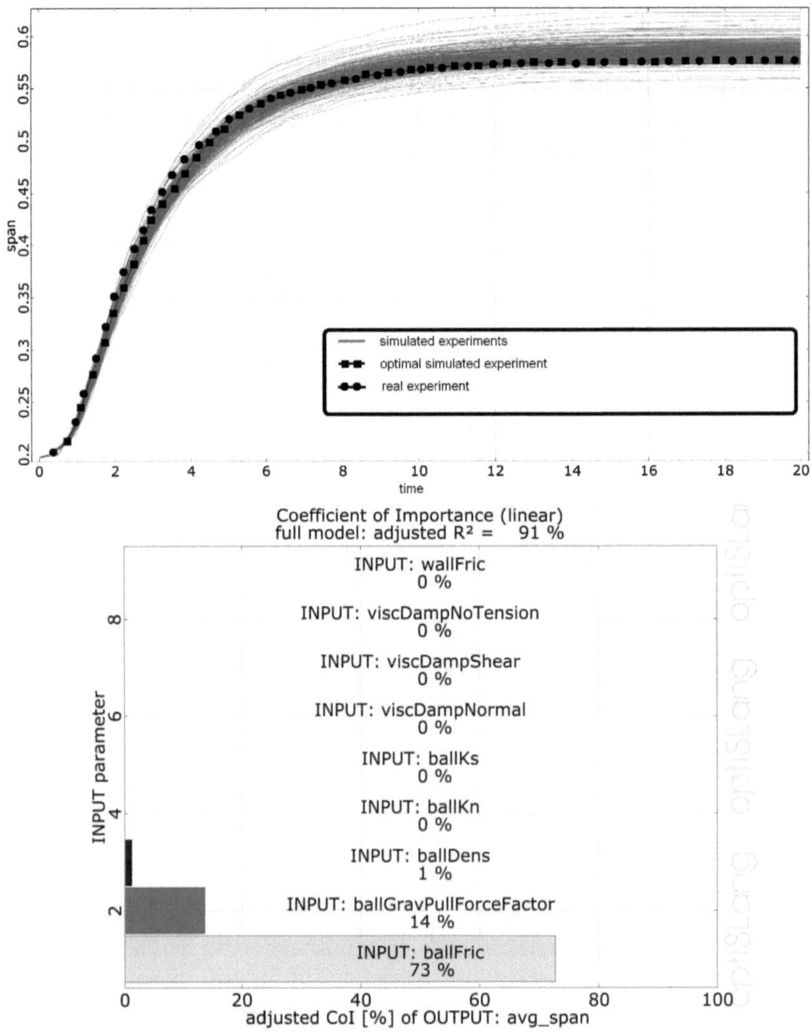

Fig. 3.10 (a) Results of slump flow over time in different simulations with a simple contact model, defined by automated calibration (grey) in comparison to the real experiment (black with dots); (b) visualization of the coefficient of importance for the dependency of simulation parameters on the results with a simple contact model [22]

An example of a granular material is shown in Fig. 3.9. Clearly visible are the different characteristics of the angles of repose and the tear-off edges. To represent the material adequately, several experiments should be modelled to capture all the relevant parameters of the material's behaviour. Additionally, the boundary conditions, such as external acceleration of the material, will vary in the reference experiments depending on the relevant ranges in the target simulation, to capture the material behaviour under different conditions. This calibration procedure is very difficult and time-consuming. Furthermore, the evaluation of the results and the adjustment of the parameters are not always accurate.

To improve this procedure there exist different approaches. One approach is an automated parameter calibration. In this case the overall procedure doesn't change, but the evaluation of results and the adjustment of parameters are processed automatically by an optimisation tool based on search heuristics with a statistical background [22]. This approach reduces the work load and the empirical influence on the result. In addition the correlations and dependencies of parameters are extracted automatically (*cf.* Fig. 3.10).

Another approach is to estimate the simulation parameters based on measured material constants, describing its behaviour according to a particular material model. A requirement for this approach is a reliable determination of dependencies between the material parameters considered, e.g., τ_0 and μ in the Bingham model, and the parameters of the numerical model. This is possible either by a complete analysis of the parameter ranges or by deriving particular parameters of the chosen contact model using specific algorithms. Both approaches are not straightforward, but there are initial successful results.

In [23] an approach aimed at establishing a link between the rheological properties of fresh concrete and the parameters of DEM-based models is presented. Generally it can be stated that the system of particles representing concrete is being plastically deformed when the tensile forces acting at the meso-level become higher than the contact bond strength. This basic principle was used first in developing the algorithm, which should provide a link between the yield stress of concrete according to the Bingham model and the bond strength parameter of the particle model. Fig. 3.11 shows this algorithm schematically.

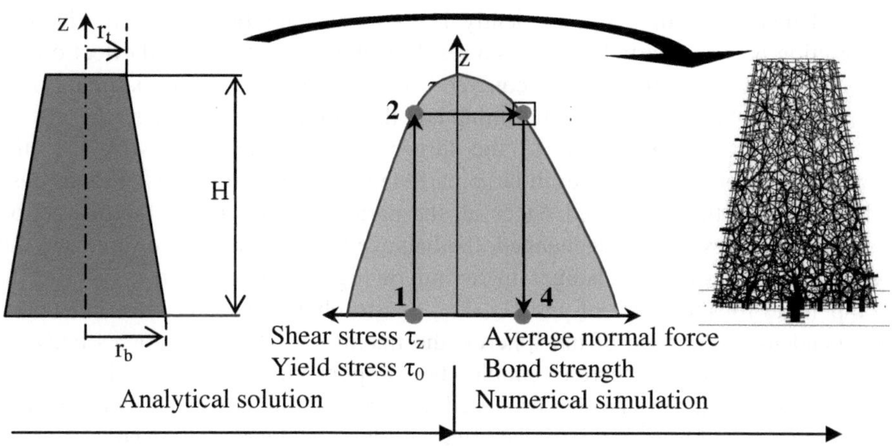

Fig. 3.11 Algorithm for the determination of bond strength [23]

The keystone of the algorithm is the comparison of the analytically predicted stress distributions in the Abram's cone filled with concrete and the corresponding force distribution obtained from a corresponding numerical simulation by means of DEM, both calculations performed before cone lifting.

3.3.5 Particle Size Effect and Dimensional Analysis

Applying the Particle Method, it is possible to simulate the effect of concrete composition on its rheological behaviour by defining the components of the concrete meso-structure, *i.e.*, fine mortar, sand particles, and coarse aggregates, using discrete particles of different sizes (*cf.* Chapter 3.3.1.). In the parameter study presented in [1] a different approach was chosen. The concrete was simulated using particles of only one size at a particular time. In this way the effects of the different model parameters are more recognisable. Such a numerical model with one-size particles can be interpreted as a multitude of round (or spherical) aggregate grains of some "average", representative size, each uniformly covered by a layer of cement paste (or fine mortar). It should be mentioned here that the assumption of a single particle size has clear advantages with regard to the prospective practical application of the numerical approach presented, since such a treatment of concrete as a collection of discrete particles is simple and the corresponding calculations are very fast.

The use of small particle sizes is limited by long calculation times arising out of the great number of particles per unit volume; the maximum size is limited by the maximum aggregate size of the concrete. Table 3.1 shows the results of the 2D and 3D simulations of the slump test using particle sizes with a radius of 2.5 mm, 3.5 mm and 5 mm, respectively, while all the other parameters of the model mentioned in the section before remained unchanged.

A change in particle size evidently leads to an alteration of the mechanical interaction between particles and, as a result of this, to a pronounced effect on the rheological behaviour of the concrete as simulated. Larger particle sizes correspond to higher values of the slump and the slump-flow diameter.

This effect can be traced to the larger particles' being heavier, *i.e.,* the gravitational force acting on each large particle is higher in comparison to the case when small particles are used. Since all the parameters describing the interaction between particles remain unchanged, the balance between the active force and the resistance to movement changes in favour of higher displacements. In order to compensate for the effect of particle size, *i.e.,* to obtain the same flow behaviour independent of the radii of the spheres, the maximum interaction forces between particles of virtual concrete should be proportional to the particle mass. Mechtcherine and Shyshko [1] proposed a corresponding correction coefficient k to compensating for the particle size effect of the same density ρ, see eq. (12).

Table 3.1 Effect of particle size on the results of the simulated slump test [1]

2D simulation		3D simulation	
Particle radius and image of the "concrete cake" at the end of simulation	Calculation data	Calculation data	Particle radius and image of the "concrete cake" at the end of simulation
5mm 	Number of particles: 3254 Slump: 12 cm Slump flow: 37 cm	Number of particles: 51551 Slump: 12 cm Slump flow: 35 cm	5 mm 
7 mm 	Number of particles: 1546 Slump: 13 cm Slump flow: 41 cm	Number of particles: 18512 Slump: 13 cm Slump flow: 37 cm	7 mm 
10 mm 	Number of particles: 624 Slump: 20 cm Slump flow: 50 cm	Number of particles: 6491 Slump: 20 cm Slump flow: 50 cm	10 mm 

$$k = \frac{m_i}{m_0} = \frac{\rho V_i}{\rho V_0} = \frac{\frac{4}{3}\pi R_i^3}{\frac{4}{3}\pi R_0^3} = \left(\frac{R_i}{R_0}\right)^3 \quad (12)$$

where m_i, V_i, and R_i are the mass, volume, and radius of particles of different size to the reference radius R_0, while m_0 and V_0 are the values for the reference particle size R_0.

A comparison of the calculated images of the concrete "cake" at the end of simulation as described in Table 1 shows that it is possible to identify trends in 2D simulations similar to those in 3D. The tremendous computational time savings for a 2D simulation, therefore, make a quick way to gather general information and trends possible. Nevertheless, to get more exact quantitative information, most processes have to be modelled in 3D to capture the effects of particle movement in all directions and to preserve the correct proportionalities of volume and surface.

3.4 Calibration and Verification

3.4.1 Slump and Slump Flow

In the numerical simulation presented in [23], concrete was modelled by particles of different sizes in such a way that a realistic grading curve of aggregates could be represented with good approximation. The particles were randomly distributed over a given volume and subsequently the cone was filled with these particles, which moved down and compacted under the force of gravity. The density of the particles was chosen such that under consideration of the average packing of particles the specific weight of the mix was equal to that of a normal-weight concrete.

Fig. 3.12a illustrates the distribution of the particles in the cone at the beginning of the calculation and Fig. 3.12b shows the contact forces acting between particles before cone lifting. The black coloured lines represent compressive forces. The thickness of these lines is proportional to the magnitude of the force. Fig. 3.12c gives the distribution of the calculated contact forces over the height of the cone. Each point represents the position (height) of a contact between two particles and the corresponding value of the contact force. The grey dashed line in the diagram connects the points corresponding to averaged maximum values of the contact force (averaged contact force for 10 points having the highest values). This curve is smoothed for the sake of clearer presentation; in reality it cannot be completely smooth because concrete is simulated in a discrete manner when using DEM.

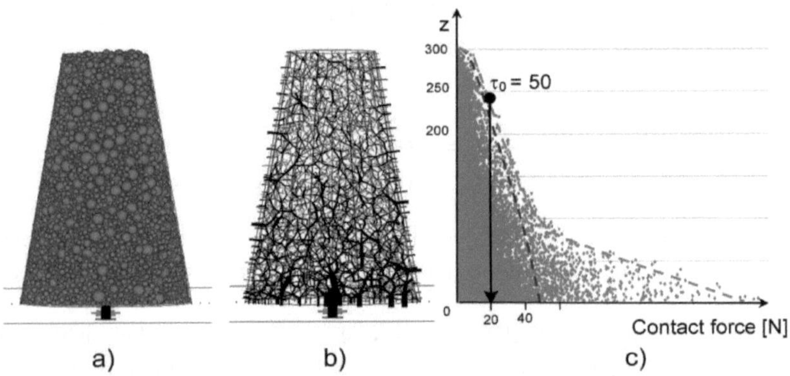

Fig. 3.12 Initial state of the numerical simulation: (a) concrete particles in the cone, (b) normal contact forces and (c) contact force distribution [23]

It should be noted that the upper two thirds of the force distribution curve have a shape that matches the analytical solution, *cf.* Fig. 3.11 (the expected shape of the curve is given in Fig. 3.12 by a black dashed line). The yield stress in this interval covers self-compacting and flowable concretes. The wide spreading of force values, especially in the lower third of the cone originates primarily from the uncompleted relaxation of the sample. Equilibrium is not yet reached at this stage of the calculation; more calculation steps would bring the result nearer to the expected force distribution, which means, however, a considerable increase in the calculation time. For the selected reference material, a self-compacting concrete with a yield stress of 50 Pa, continuing the calculation would not produce any improvement since the shape of the upper two thirds of the curves, *i.e.* the region corresponding to relatively low values of yield stress, would not change.

The slump-flow value obtained from the simulation was 580 mm. The analytical prediction using the formula by Roussel and Coussot [24] provides a value of 600 mm with respect to an input yield stress of 50 Pa. Such a good correspondence of numerical and analytical results can be regarded as a first validation of the methodology developed to derive the key parameter of the particle model, the bond strength.

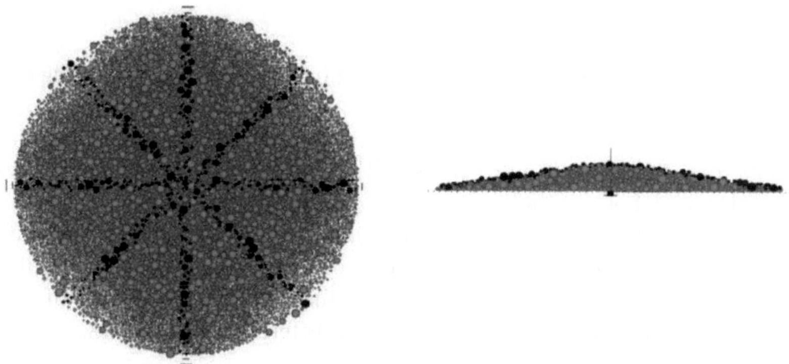

Fig. 3.13 Final shape of the concrete cake in slump flow test with PFC3D [23]

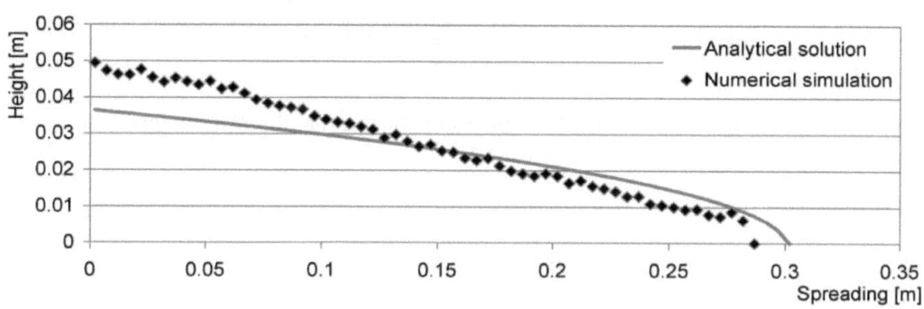

Fig. 3.14 Results of the numerical and analytical prediction of the final shape of concrete in a slump flow test for the yield stress of 50 Pa [23] (analytical solution according to [24])

In order to obtain the height profile of the virtual concrete cake, 8 directions, with an interval of 45°, were considered, *cf.* the black stripes in Fig. 3.13. The average values (average of the maximal heights in each direction in same radial interval) in radial direction are presented in Fig. 3.14. For comparison the analytical solution by Roussel and Coussot [24] is given.

Fig. 3.15 shows the calculated final shape of the concrete cake after slump flow test and the corresponding result of the actual experiment on SCC. The numerical simulation provides a good description of material behaviour as observed in experiment.

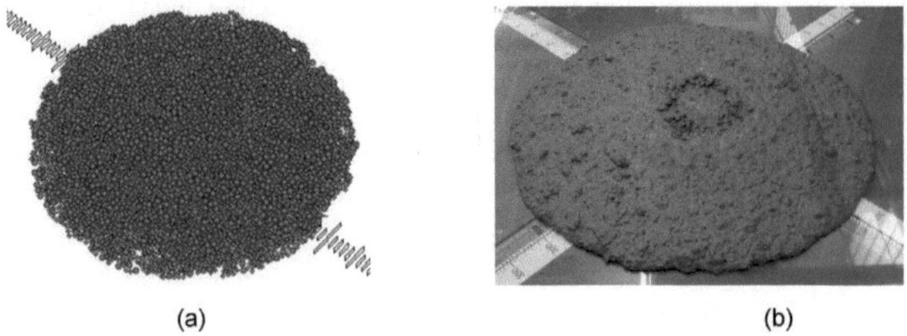

Fig. 3.15 Final shape of concrete as obtained from (a) numerical simulation using polysized particles and (b) slump flow test with SCC [1]

Simulations with large (5 mm) and small (2.5 mm) particle sizes as well as with a composition of both particle sizes were processed in EDEM, to illustrate the effects of different particle sizes on the final shape of the concrete cake. In Fig. 3.16 the three-dimensional front view of the simulation result with small particle size is displayed. Fig. 3.17 presents further the comparison of the final height distribution of this slump flow test with different particle sizes.

Fig. 3.16 Final stage of the EDEM simulation showing the slump flow test with particle size of 2.5; simulation by the IFF Weimar e.V.

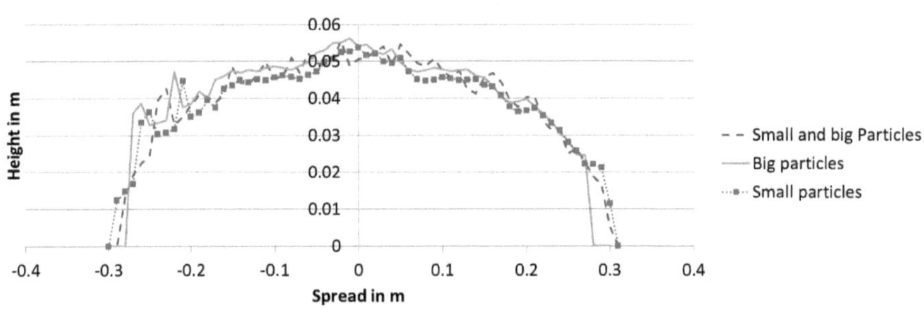

Fig. 3.17 Height distribution of slump flow simulation in EDEM with three different particle sizes (picture courtesy of IFF Weimar e.V.)

In [9] the slump flow test, based upon ASTM C 143, was used to simulate workability of fresh concrete with different consistencies, both with and without fibres. Fig. 3.18 shows the result of the simulation of a slump flow test on fine-grained numerical concrete with the workability of ordinary concrete, a self-compacting concrete (SCC), and of a self-compacting concrete with steel fibre reinforcement, respectively. Their characteristic behaviour profiles, based upon experimental results of the corresponding types of concrete mentioned, were attainable by appropriate choice of the model parameters. Furthermore, the effect of fibre addition to SCC on slump flow behaviour could be shown correctly.

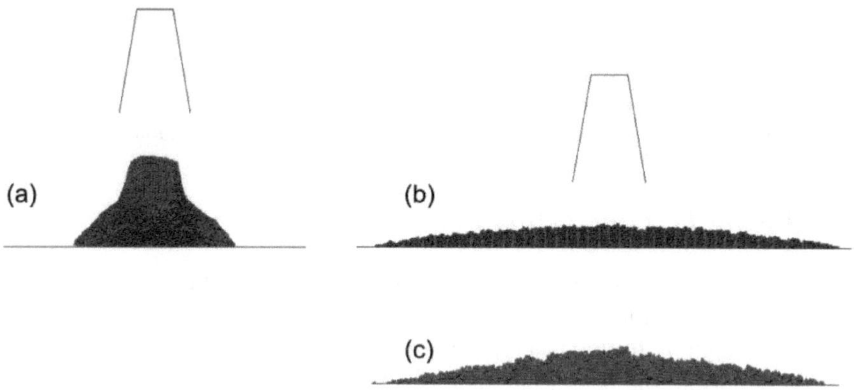

Fig. 3.18 2D simulation of slump flow tests on (a) ordinary concrete, (b) SCC and (c) SCC with 1% fibre (by volume) [9]

After calibrating the fine-grained model concrete in conjunction with the workability of an ordinary concrete, the effect of adding coarse aggregates and fibre to the mixture was studied. According to Fig. 3.19a such a modification to concrete composition, without any change in the model, clearly leads to stiffer behaviour of the fresh concrete, a finding which is in agreement with experimental observations. Obviously, the addition of fibre resulted – due to the distinct slenderness of fibre – in a more pronounced change of stiffness of concrete in comparison to the case when a considerably higher portion of coarse aggregates was added.

Furthermore, the compaction process of concrete was simulated by oscillation of the wall element, which represented the top plate of a spreading table. This procedure simulated the compacting work induced by shocking the concrete in accordance with the regulations outlined in the Code EN 12350-5:2009 (Testing fresh concrete - Part 5: Flow table test). Fig. 3.19b is an example displaying the result of such compaction for fine-grained fibre reinforced concrete. A rather realistic response of the simulated material was observed.

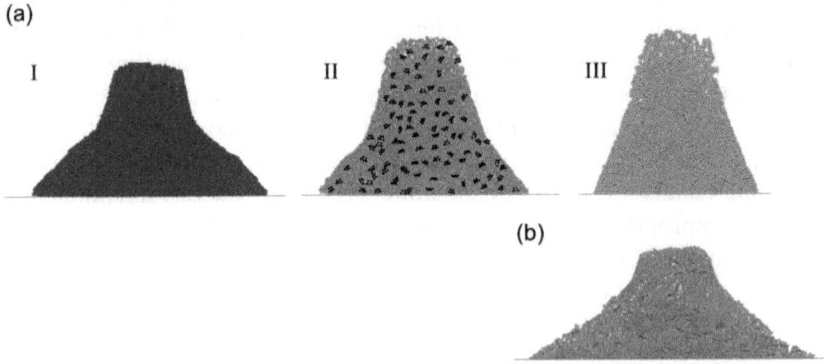

Fig. 3.19 (a) Simulation of slump-flow tests on ordinary concrete: I. fine-grained concrete, II. concrete with coarse aggregates, and III. fibre-reinforced fine-grained concrete, (b) Fibre-reinforced concrete after compaction by shocking on the spreading table [9]

3.4.2 J-Ring Test and L-Box Test

The slump-flow test usually serves as the basis for parameter studies as well as for model calibration. Additionally, in order to assess the predictive ability of the numerical simulation, further test types have to be used as reference to validate the model further while the model parameters are held constant. These two tests are common in the practice of testing SCC: the J-Ring and L-Box tests.

3.4.2.1 J-Ring Test

The J-Ring test is used to assess the ability of SCC to flow around rebar in reinforced concrete structures. It shows whether or not segregation and blocking of coarse aggregate grains occur at steel bars when SCC passes through the spacing between them. Fig. 3.20 presents the final state in the numerical simulation of this test obtained by using polysize particles and, as a comparison, the result of the corresponding experiment. Similar to the experimental result, a higher concrete level was observed inside than outside the steel ring in the numerical simulation. The diameter of the maximum spreading was nearly the same in the experiment and in the simulation (it was equal to 55 cm). It should be emphasised that no parameter fine-tuning occurred here. The parameters were simply taken from the calibration procedure with the corresponding slump-flow test [1].

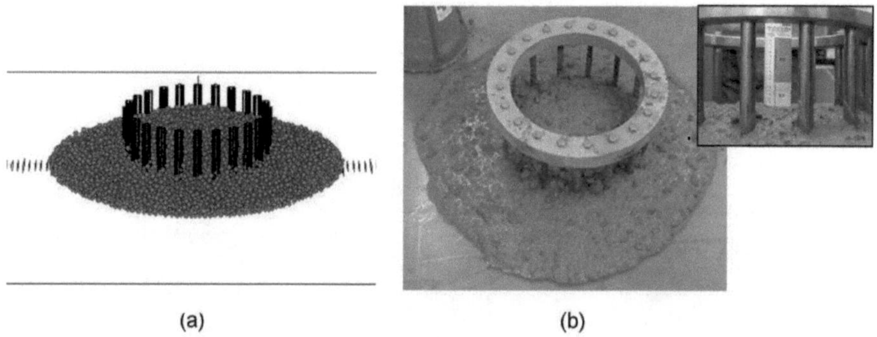

Fig. 3.20 (a) Numerical and (b) experimental results of a J-Ring test with SCC [1]

3.4.2.2 L-Box Test

The L-Box is commonly used to assess the filling ability and passing ability or, in another view, the blocking behaviour of SCC. Fig. 3.21 presents simulations for the characterisation of the flow behaviour in both the L-box and the Slump Flow tests (Abrams cone). Note that the simulated concrete aggregates are evenly spread in the slump, but have been temporarily held back a bit behind the bars of the L-box [14].

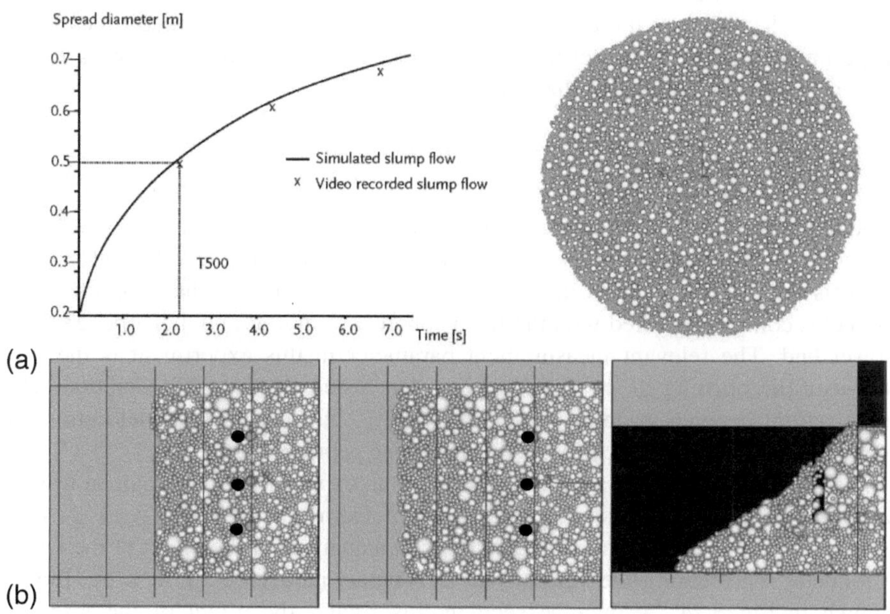

Fig. 3.21 (a) The slump flow of simulated concrete, compared to a video recorded slump flow; (b) computer simulation results for the L-Box (views from above and side view of the L-Box) [14]

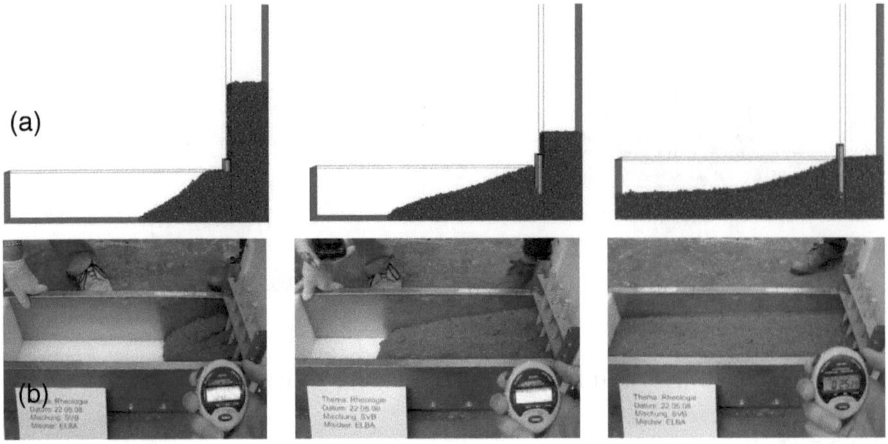

Fig. 3.22 Results of (a) the numerical simulation and of (b) the L-Box test for three different times (when the flow front reaches the marks 20cm and 40cm and the final state) [1]

Fig. 3.22 also shows three snapshots from the numerical simulation. They were made when flow front reached the 20 cm mark, 40 cm mark, and the final state. The same figure presents the corresponding images from the L-Box test on SCC. The results of the simulation and experiment are in satisfactory agreement with regard to the shape of the concrete flow. Furthermore, the measured times in the experiment (when these defined states were reached) correspond well to the internal computation time of the DEM program [1].

3.4.3 Funnel Flow

The funnel flow test is another common procedure to evaluate the workability and rheology of SCC and mortars. Depending on the material mix, different sizes of funnels are used as well as different shapes. In carrying out the experiment the funnel is completely filled with the fresh concrete or mortar and then opened at the lower end. The relevant measurement parameter in this experiment is the time between the opening of the gate and the state when the funnel is emptied. This time period depends primarily on the viscosity, but also on the yield stress. A numerical solution of the cylindrical marsh cone is presented in [25].

The funnel flow test is another example of an experiment which can be used as a validation test for the simulation model of fresh concrete. Fig. 3.23 shows a simulation of the process at different stages. An analysis of the effect of the DEM-parameter representing the viscosity in the contact model on the flow out time is shown in Fig. 3.24. As can be seen in this figure, the effect of the viscosity representing parameter in the simulation is the relevant factor of influence for the flow time.

Fig. 3.23 Simulation of the funnel flow (picture with the courtesy of IFF Weimar e.V.)

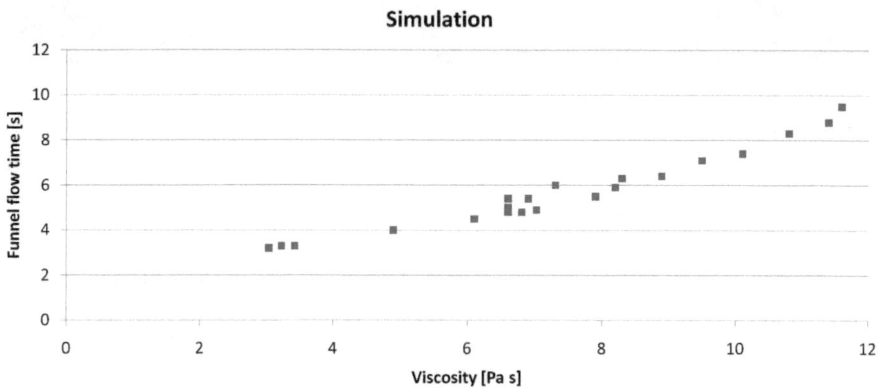

Fig. 3.24 Dependency of the parameter representing the viscosity on the funnel flow time in the simulation (picture courtesy of IFF Weimar e.V.)

3.4.4 Casting

After calibrating a numerical concrete model by means of the simulation of slump flow tests, a series of filling process simulations of a beam mould measuring 500 mm by 100 mm by 100 mm was performed by the group at the TU Dresden [9]. In this series, a numerical SCC was considered both with and without fibres. Following the procedure in the real SCC handling, no compaction work was introduced in the simulation. Furthermore, a mixture with the rheological properties of ordinary concrete was considered. In this case, a compaction by means of vibration was simulated by the oscillation of the wall elements.

Fig. 3.25 illustrates one of the steps from the 2D-simulation of the mould's being filled with fibre reinforced self-compacting concrete. The particles, which represent the concrete, are coloured light grey, while the fibres are distinguished with a black colour. It can be recognised that the fibres' preferred orientation is in the direction of the flowing of concrete.

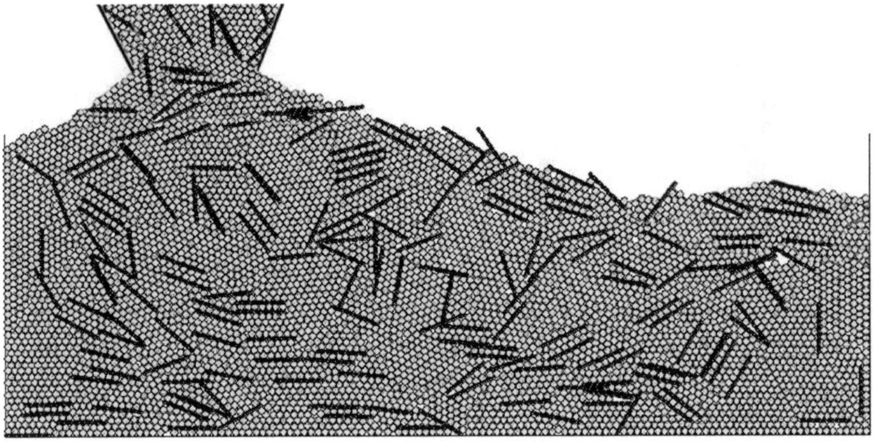

Fig. 3.25 Simulation of mould filling with fibre-reinforced SCC [9]

3.5 Industrial Applications

The DEM can be used to model several fields of industrial applications for fresh concrete, *e.g.*, mixing, filling, extrusion, transport, and compaction. Depending on the consistency of the fresh concrete, the adequate contact model has to be applied within the simulation. The focus of the material models in industrial applications is on a realistic material behaviour as well as on fast computing. There are different goals for using DEM to model fresh concrete processing. On one hand it can be useful just to gain insight into the process of interest in order to get more information about it. Another aspect is the optimisation of machine layout to improve the process. The process improvement thereby can mean shortening the processing time to increase product quality or lower costs. Simulating different machine layouts is an energy-efficient, resource-saving alternative to prototyping, and as such it can save time and expense. Besides the testing of machine layouts, process parameters like rotational speeds, processing time and so on can be varied easily within the simulation to test the influence on the process and its particular results.

3.5.1 Mixing

The goal of the concrete mixing process is to produce a homogeneous mixture of the separately poured grain fractions. One of the inherent problems here consists in evaluating the mixing quality. Because of the non-transparent cement suspension, visual inspection is not possible in the experiment, and samples of fresh concrete have to be taken, washed out, and sieved in order to estimate the homogeneity of the mixture. The task of simulation is to model the movement of the mixture within the mixer and to analyse the blending of the grain ingredients. The objective thereby is to find a design of the mixing tools and the movement of these instruments to reach the optimal mixture quality or to reach a defined mixture quality as rapidly and with as much energy efficiency as possible. Besides the general benefits of simulation there are additional advantages in using this technique to optimise such mixing:

- the easy evaluation of mixing quality at any time without interrupting the mixing process,
- the "transparent" view enabling detection of dead zones.

As an example for the application of DEM on mixing processes, Fig. 3.26 and Fig. 3.27 show the simulation of a planetary mixer.

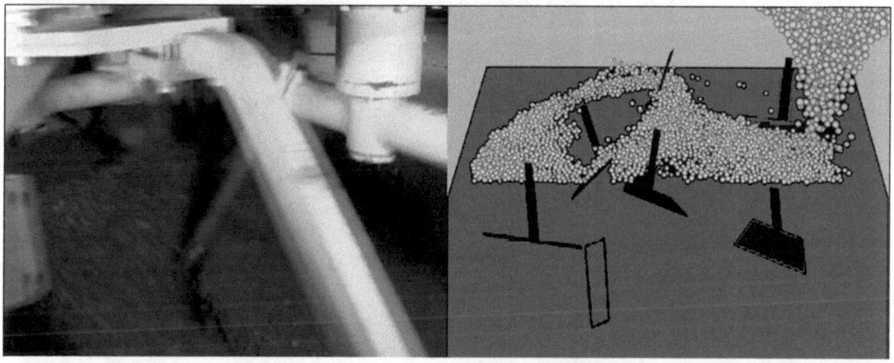

Fig. 3.26 Comparison of a real mixer and a DEM simulation (figure courtesy of IFF Weimar e.V.)

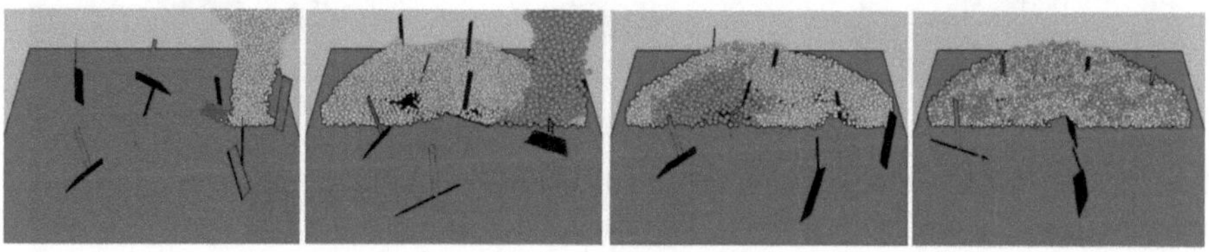

Fig. 3.27 Insight view of a DEM simulation of a planetary mixer at different mixing stages. Different colours (grey scale) represent different grain sizes (figure courtesy of IFF Weimar e.V.).

3.5.2 Filling

The filling process has a very wide range of application, starting with small volumes like moulds of stone block machines operating with very stiff concrete mixtures up to huge, heavily reinforced formworks to be filled with SCC. The boundary conditions differ depending on the individual process, but the goal is the same: a homogeneous and complete filling of the given volume. Specific problems like segregation, blocking, and limited filling time are task-dependent. To bring a better prediction of filling behaviour to light, DEM simulation is used.

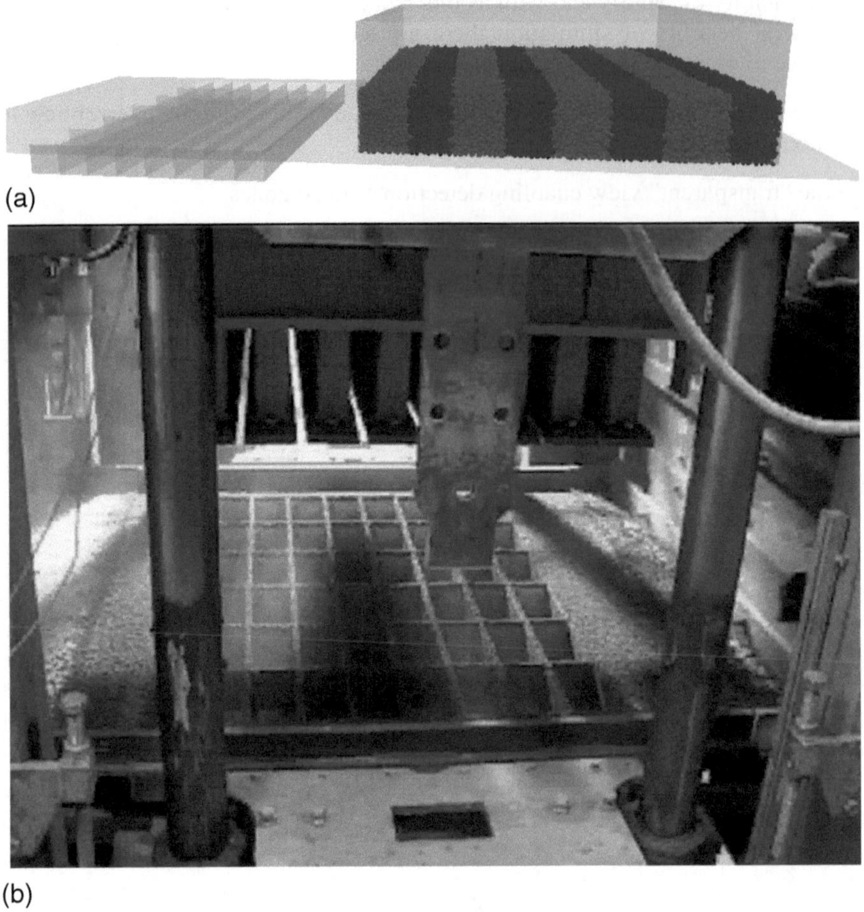

Fig. 3.28 (a) A DEM discretisation for simulating the mould filling process and (b) a real filling wagon, (figures courtesy of IFF Weimar e.V.)

The filling of huge formworks is nearly impossible to model because of its high demands on computation time; so, smaller volumes should be used to represent the critical sections. Apart from that, applications like the filling of stone block machine moulds with smaller volumes are well realisable in DEM. In this special case the target is to fill each cell of the mould completely and as fast as possible because the shorter the filling process time, the more stones can be produced. Incorrectly filled mould cells will produce stone of inferior quality. Moreover, unevenly distributed filling over the cell mould can lead locally to different compaction intensities and subsequently give rise to unacceptable product quality in some stones. A long resting time of the filling wagon above the mould cells can guarantee an appropriate filling level but will increase the filling time. Thus, there are competing goals to be reached. To improve the filling level without increasing the time, special geometries and moving regimes of the filling wagon are necessary. Because the filling wagon is included into the block stone machine, it is difficult in reality to test wagon layouts differing from the normal design. Therefore, simulation in this case is a very promising solution strategy. An example of simulating the filling process is shown in Fig. 3.28 and Fig. 3.29.

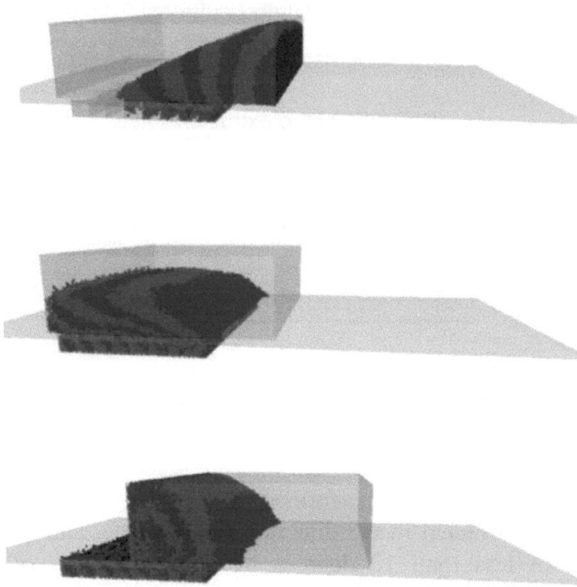

Fig. 3.29 DEM simulation of a mould filling process (the different particle colours are used to improve the visibility of the movement), (figure courtesy of IFF Weimar e.V.)

3.5.3 Extrusion

Extrusion is a manufacturing process used to produce long objects of a fixed cross-sectional profile. This process is also used for production of elements of mortar or concrete, often reinforced by fibre. The mechanical performance of the fibre-reinforced concrete (FRC) or mortar depends among others on the fibre alignment. High shear forces occurring during the forming process forces short fibres to be oriented in the direction of the extrusion, see, for example, [26]. Such fibre alignment improves the mechanical performance of FRC in the extrusion, which is beneficial for components that are designed to carry tensile load in only one direction.

To achieve optimum fibre alignment and distribution and at the same time obviate extremely high shear forces in the equipment and fibre segregation, the rheological behaviour of the concrete as well as the geometry of the equipment, velocity of the extrusion, and other parameters must be optimised. Here again, the numerically simulated extrusion provides an inexpensive method in dealing with this task. The advantage of the simulation is the possibility of visualising the flow of the fibres and tracking their positions and orientation during the entire extrusion process. Furthermore, there is the possibility of easy modification of the initial simulation conditions such as geometry of the equipment and parameters of the extrusion process.

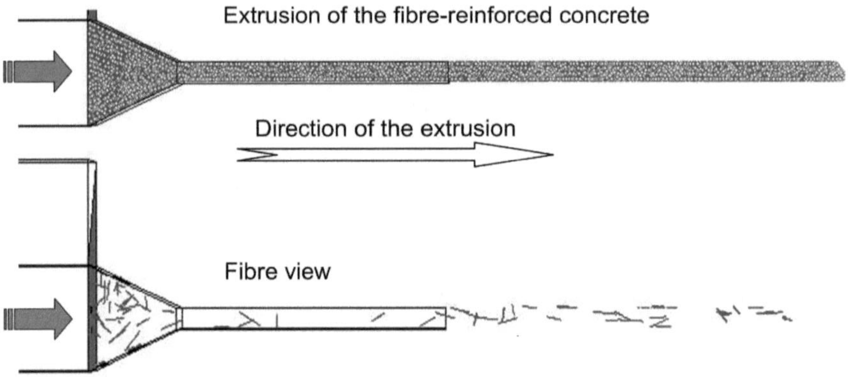

Fig. 3.30 (a) Extrusion of the fibre-reinforced mortar; (b) view of the fibre distribution in the mortar (figure courtesy of TU Dresden)

The results of the simulations performed by the group at the TU Dresden show 90% of the fibres reoriented in the direction of extrusion, and 60% of the fibres show an angle of less than 10 degrees from the direction of extrusion, see Fig. 3.30.

3.6 Future Perspectives

In future one possible improvement is the combination of DEM techniques with CFD. In this case there are solid and fluid phases which interact bidirectionally. As SCC is a suspension with discrete particles, this approach seems to be very promising. So far there are several problems to be solved:

- high computation time for coupled simulations of CFD and DEM,
- adequate implementation of coupling for a very high particle fraction.

As calibration is still a significant problem concerning effort and accuracy, an improved and simplified way to adjust the simulation parameters is the aim. This is to reduce the effort in starting the simulation of complex processes. Even though there are first approaches toward reaching this goal, there is still much potential for improvement.

As the computational power of CPUs is still steadily rising and algorithms are being optimised, the DEM will be able to simulate ever smaller particles. This will enable a sieving grade in the simulation that is closer to reality. Hence, DEM will be able to model more effects depending on the size distribution directly and produce more accurate results.

Further research is also needed with regard to the simulation of de-airing processes of fresh concrete.

An advantage of DEM is that this approach is universal and can cover very different processes and phenomena. As was shown in [8], DEM-based models can enable the user to analyse individual processes at different stages of concrete life, including specific transitions from one state into another. Such an approach clearly implies that some compromises are necessary since the particular general model might not provide the best possible technique for every individual process. However, the benefits of such continuous modelling, which involves the development of concrete properties in time under consideration of changing exposures, might prevail in many cases.

The ultimate aim of such an approach – a kind of virtual concrete laboratory – is demonstrated in Fig. 3.31. In the example given, three stages of production and testing of a concrete beam made of fibre reinforced concrete are shown. After mixing, depending on the rheological behaviour of the concrete, its fibre distribution, its orientation, its de-airing and, possibly, its segregation are influenced by the process of filling the mould and concrete compaction. Eventually, the quality of the concrete's de-airing, fibre orientation and degree of homogeneity affect the mechanical fracture behaviour of concrete specimens in three-point bend tests.

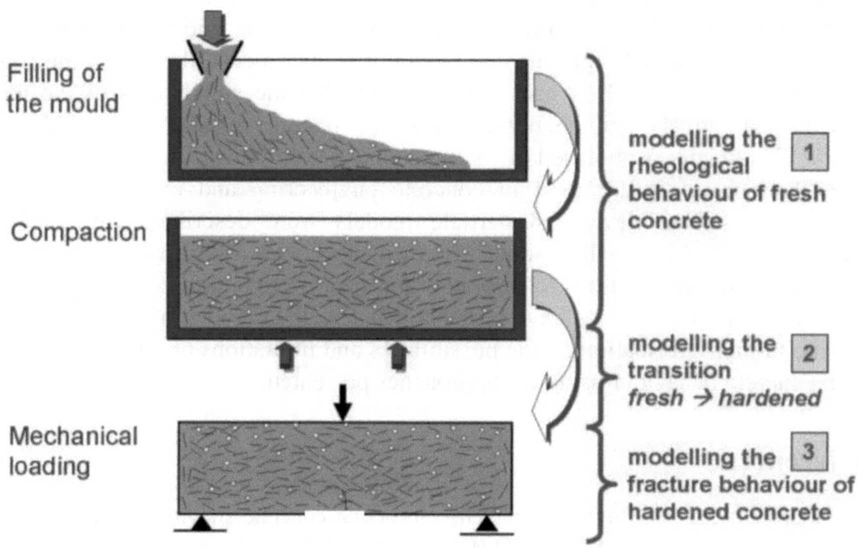

Fig. 3.31 Schematic view of the individual stages of the concrete life; here an example of the specimen production and testing [8]

The possible spectrum of relevant processes which might be incorporated into the model is expected to be very broad. For example, with regard to the durability and the transport of fluids and gases through the cracks induced by mechanical loading or shrinkage can be simulated using DEM. Such transport of aggressive substances is crucial to the forecasting of concrete deterioration. Also, the deterioration of concrete due to rebar corrosion, abrasion, *etc.* can be modelled. The simulation of the process describing the transition of concrete from the fresh state to hardened state, *i.e.*, the hardening process with all the accompanying time-dependent phenomena, such as development of strength and stiffness, autogenous shrinkage, development of internal stresses induced by the shrinkage of the cement paste, concrete creep, *etc.*, is a subject of ongoing investigation. Some basic considerations are given in [8].

3.7 Summary

This chapter provides an overview of the development and the contemporary state of research in the field of simulating fresh concrete flow using the Discrete Element Method (DEM). First, the mathematical methodology and the modelling of rheological behaviour were explained followed by the identification of the model parameters. Particular emphasis was given to the approaches aimed at establishing a link between the rheological properties of fresh concrete and the parameters of DEM-based models. Various examples of the estimation of model parameters and calibration of the model were demonstrated. Subsequently, the quality of such parameter estimation was verified by comparing the final shapes of the concrete cakes obtained from the numerical simulations and the corresponding predictions by analytical formula and laboratory experiments like the Slump Flow test, J-Ring test and L-Box test. It was shown that the simulations of the tests provide qualitatively and quantitatively sound results, displaying correctly the critical phenomena as observed in corresponding experiments.

Furthermore, software used in concrete engineering and existing industrial applications of the developed particle models were described, showing the potential of DEM. Finally, some representative and promising examples of industrial applications of this relative new approach were presented.

Some future perspectives have been shown as well. Ongoing investigations will provide further information on the possibilities and limitations of the simulation of fresh concrete behaviour using the approaches presented.

References

[1] Mechtcherine, V., Shyshko, S.: Self-compacting concrete simulation using Distinct Element Method. In: Wallevik, O.H., Kubens, S., Oesterheld, S. (eds.) Proceedings of the 3rd International RILEM Symposium on Rheology of Cement Suspensions such as Fresh Concrete, Reykjavik, Iceland, August 19-21, pp. 171–179. RILEM Publications, Bagneux (2009)

[2] Kishino, Y.: Powders and Grains 2001. In: Proceedings of the Fourth Interntional Conference on Micromechanics of Granular Media, Sendaï, Japan, May 21-25, A.A. Balkema Publishers, Lisse (2001)
[3] Chu, H., Machida, A., Suzuki, N.: Experimental investigation and DEM simulation of filling capacity of fresh concrete. Transactions of the Japan Concrete Institute 16, 9–14 (1996)
[4] Noor, M.A., Uomoto, T.: Three-dimensional discrete element simulation of rheology tests of Self-Compacting Concrete. In: Skarendahl, Å., Petersson, Ö. (eds.) Proceedings of the 1st International RILEM Symposium on Self-Compacting Concrete, Stockholm, Sweden, September 13-14, pp. 35–46. RILEM Publications, Cachan (1999)
[5] Petersson, Ö., Hakami, H.: Simulation of SCC – laboratory experiments and numerical modeling of slump flow and L-box tests. In: Ozawa, K., Ouchi, M. (eds.) Proceedings of the 2nd International Symposium on Self-Compacting Concrete, Tokyo, Japan, October 23-25, pp. 79–88. Coms Engineering Corporation, Tokyo (2001)
[6] Petersson, Ö.: Simulation of Self-Compacting Concrete – laboratory experiments and numerical modeling of testing methods, J-ring and L-box tests. In: Wallevik, Ó., Níelsson, I. (eds.) Proceedings of the 3rd International RILEM Symposium on Self-compacting Concrete, Reykjavik, Iceland, August 17-20, pp. 202–207. RILEM Publications, Bagneux (2003)
[7] Shyshko, S., Mechtcherine, V.: Continuous numerical modelling of concrete from fresh to hardened state. In: Finger-Institut für Baustoffkunde, F.A. (ed.) Tagungsbericht der 16. Internationalen Baustofftagung, Ibausil, Weimar, Germany, September 20-23, vol. 2, pp. 165–172. Bauhaus Universität, Weimar (2006)
[8] Mechtcherine, V., Shyshko, S.: Virtual concrete laboratory – Continuous numerical simulation of concrete behaviour from fresh to hardened state. In: Grosse, C.U. (ed.) Advances in Construction Materials, pp. 479–488. Springer, Berlin (2007)
[9] Mechtcherine, V., Shyshko, S.: Simulating the behaviour of fresh concrete using Distinct Element Method. In: De Schutter, G., Boel, V. (eds.) Proceedings of the 5th International RILEM Symposium on Self-Compacting Concrete, SCC 2007, Ghent, Belgium, September 3-5, pp. 467–472. RILEM Publications, Bagneux (2007)
[10] Kuch, H., Palzer, S., Schwabe, J.-H.: Anwendung der Simulation bei der Verarbeitung von Gemengen. In: Finger-Institut für Baustoffkunde, F.A. (ed.) Tagungsbericht der 16. Internationalen Baustofftagung, Ibausil, Weimar, Germany, September 20-23, vol. 1, pp. 1321–1327. Bauhaus Universität, Weimar (2006)
[11] Schwabe, J.-H., Kuch, H.: Development and control of concrete mix processing procedures. In: Borghoff, M., Gottschalg, A., Mehl, R. (eds.) Proceedings of the 18th BIBM International Congress and Exhibition, Amsterdam, the Netherlands, Amsterdam, the Netherlands, May 11-14, pp. 108–109. Bond van Fabrikanten van Betonproducten in Nederland, Woerden (2005)
[12] Numerical Modelling in Micromechanics via Particle Methods. In: Konietzky, H. (ed.) Proceedings of the 1st International PFC Symposium, Gelsenkirchen, Germany, November 6-8, A.A.Balkema Publishers, Lisse (2002)
[13] Shyshko, S., Mechtcherine, V.: Simulating the workability of fresh concrete. In: Schlangen, E., De Schutter, G. (eds.) Proceedings of the International RILEM Symposium of Concrete Modelling, CONMOD 2008, Delft, the Netherlands, May 26-28, pp. 173–181 (2008)
[14] Itasca Consulting Group Inc.: PFC 2D. Version 3.0. ICG, Minneapolis (2002)

[15] Gram, A., Silfwerbrand, J.: Numerical simulation of fresh SCC flow: applications. Materials and Structures, vol. 44, pp. 805–813 (2011)

[16] Gram, A.: Numerical Modelling of Self-Compacting Concrete Flow - Discrete and Continuous Approach. Licentiate thesis, Royal Institute of Technology, Stockholm (2009)

[17] Chu, H., Machida, A.: Numerical simulation of fluidity behaviour of fresh concrete by 2D distinct element method. Transactions of the Japan Concrete Institute 18, 1–8 (1996)

[18] Cundall, P.A., Konietzky, H., Potyondy, D.O.: PFC – Ein Neues Werkzeug Für Numerische Modellierungen. Bautechnik 73(8), 492–498 (1996)

[19] Malkin, A.Y., Isayev, A.I.: Rheology - Concepts, methods & applications. ChemTec Publishing, Toronto (2006)

[20] Macosko, C.W.: Rheology principles, measurements and applications. Wiley-VCH, New York (1994)

[21] Roussel, N.: Three-dimensional numerical simulations of slump tests. Annual Transactions of the Nordic Rheology Society 12, 55–62 (2004)

[22] Krenzer, K., Schwabe, J.-H.: Calibration of parameters for particle simulation of building materials, using stochastic optimization procedures. In: Wallevik, O.H., Kubens, S., Oesterheld, S. (eds.) Proceedings of the 3rd International RILEM Symposium on Rheology of Cement Suspensions such as Fresh Concrete, Reykjavik, Iceland, August 19-21, RILEM Publications, Bagneux (2009)

[23] Shyshko, S., Mechtcherine, V.: Simulating fresh concrete behaviour – Establishing a link between the Bingham model and parameters of a DEM-based numerical model. In: Brameshuber, W. (ed.) HetMat – Modelling of Heteroginous Materials. RILEM Proceedings PRO, vol. 76, pp. 211–219. RILEM Publications S.A.R.L. (2010)

[24] Roussel, N., Coussot, P.: Fifty-cent rheometer for yield stress measurements - from slump to spreading flow. J. Rheo. 49(3), 705–718 (2005)

[25] Roussel, N., Le Roy, R.: Marsh cone: a test or a rheological apparatus? Cement and Concrete Research 35(5), 823–830 (2005)

[26] Takashima, H., Miyagai, K., Hashida, T., Li, V.C.: A design approach for the mechanical properties of polypropylene discontinuous fiber reinforced cementitious composites by extrusion molding. Eng. Fract. Mech. 70(7-8), 853–870 (2003), doi:10.1016/S0013-7944(02)00154-6

Chapter 4
Numerical Errors in CFD and DEM Modeling

Jon Elvar Wallevik[1,*], Knut Krenzer[2], and Jörg-Henry Schwabe[3]

[1] Innovation Center Iceland, ICI Rheocenter, Iceland
[2] Institut für Fertigteiltechnik und Fertigbau Weimar e.V., Germany
[3] University of Applied Science Jena, Germany

4.1 Introduction

Numerical error present in Computational Fluid Dynamics (CFD) is given in Chapters 4.1. to 4.6. In the last chapter of this document, Chapter 4.7, a special attention is given to the error present for the Discrete Element Method (DEM). However, it should be clear that much of the topic present in Chapters 4.1 to 4.6 applies also for DEM, and other numerical flow techniques not mentioned here.

The theme "numerical error in CFD and DEM" is large and a whole science in itself. As such, the current text can only serve as an introductory text about the subject, in which the given references can provide additional information. Understanding numerical error is of paramount importance. As the availability of both CFD and DEM softwares and necessary hardware is increasing, more and more unskilled personals will start to use this tool. Bad use of CFD and DEM is for example done by incorrect use of the particular software, lack of understanding in the mathematics, accepting a particular flow result without criticism,... and the list goes on.

The current text is aimed to introduce newcomers, and in part to more experienced users, to the science of numerical error. The aim is to explain why an apparently good CFD and DEM flow results may/can be completely incorrect. The aim is to keep this text as simple as possible, but because of the nature of this theme, some mathematics is necessary.

The consequences of inaccurate flow results is in the best case wasted time, money and effort and at the worst case catastrophic failure terms of utilizing CFD and DEM as a casting prediction tool or flow analysis tool for example.

Now, the attention will begin with the CFD error, but as already mentioned above, much of this topic applies equally to DEM (see Chapter 4.7 for specific DEM error contribution).

The application of CFD modeling as a tool for the fresh concrete science can only be justified on the basis on its accuracy and the level of confidence in its results. There exist several guidelines for the best practice in CFD. The most

* Corresponding author.

N. Roussel and A. Gram (eds.), *Simulation of Fresh Concrete Flow*, RILEM State-of-the-Art Reports 15,
DOI: 10.1007/978-94-017-8884-7_4,

prominent ones are the AIAA and ERCOFTAC guidelines [1,2,3]. These guidelines set out the main rules for the conduct of CFD modeling studies. The AIAA guidelines give a particular emphasis on very complex systems, while the ERCOFTAC give best practice rules for the conduct of less complex flow problems. For the latter, the focus is on the prediction of single phase fluid flow and heat transfer and methods to quantify and minimize all sources of error and uncertainty. Thus, the ERCOFTAC manual can be considered more relevant for casting predictions of fresh concrete (in formwork and mold). In the context of trust and confidence in CFD modeling, the following definitions of error and uncertainty have now been widely accepted [1,4].

The main causes of error are

- (**E1**) NUMERICAL ERRORS: Discretization error, iterative convergence error, round off error.
- (**E2**) CODING ERROR: Error generated from source coding mistakes or bugs in the software.
- (**E3**) USER ERROR: Error generated by the operator of the software, due to his/hers lack of knowledge in CFD as well as in incorrect use of the particular CFD software being used.

The main sources of uncertainty are:

- (**U1**) INPUT UNCERTAINTY: Inaccuracies due to limited information or approximate representation of geometry, boundary condition, material properties and so forth.
- (**U2**) PHYSICAL MODEL UNCERTAINTY: Inconsistency between real flows and CFD due to inadequate representation of physical processes or due to simplifying assumption in the modeling process (for example Bingham vs. the Herschel-Bulkley, the use of regularization approach and so forth).

Before going into the details of numerical errors, some understanding about the basics of CFD is necessary and this is made in the next section. In general terms, the topic has without doubt already been covered elsewhere in TC 222. However, the CFD introduction made below is with numerical error in mind and their explanations.

It should be clear that many of the error presented here, do not only apply for CFD, but also for other types of flow calculations, like the Discrete Element Method (DEM). Thus the user of such method will benefit to read the following chapter, to realize the possible sources of errors in numerical calculations of any kind.

4.2 Basics of CFD – Understanding the Source of Errors

In its most simplistic form, CFD consist of calculating the flow by solving the equation of motion (or the Cauchy equation of motion)

$$\rho\left(\frac{\partial \mathbf{v}}{\partial t}+\mathbf{v}\cdot\nabla\mathbf{v}\right)=\nabla\cdot\boldsymbol{\sigma}+\rho\,\mathbf{g} \tag{1}$$

the term **v** being the 4 dimensional (time being one dimension) and 3 directional velocity, ρ is the density, t is the time, **g** is the gravity and $\boldsymbol{\sigma}$ is the (total) stress tensor [5,6]. The designation *Navier-Stokes* for the above equation is not preferable as it would imply a constant shear viscosity (*i.e.* constant apparent viscosity) [5,6] and thus a flow for Newtonian fluid, which is not the case for the flow of cement based material. The stress tensor $\boldsymbol{\sigma}$ will be treated later in Section 4.6.2, then in relation of computational implementation of viscoplastic material. Eq. (1) constitutes something called a partial differential equation, or PDE. Including Eq. (1), the continuity equation [5,6] is also often included in a CFD problem (especially with the presence of pressure gradient, *cf.* SIMPLE and others). But to keep the text simple and within the scope of this work the continuity equation is not included in the current discussion (for compressible fluid, the energy equation is also coupled with Eq. (1), but as the fresh concrete is incompressible, such discussion is neither needed).

In basic terms, the first and fundamental step in CFD consists of converting the equation of motion Eq. (1) into a system of algebraic equations. The latter is then programmed (in FORTRAN, C, C++, PASCAL or something similar) and thereafter converted to executables (by a compiler), which is an instruction file that a computer can understand. These steps are necessary, because in fundamental terms, a computer cannot understand or grasp a PDE in its raw form like shown with Eq. (1).

Below, we will in the absolute simplest terms explain how Eq. (1) is converted into system of algebraic equations. This will be most beneficial when discussing the sources of CFD errors. However, before this, the Taylor approximation must be discussed.

4.2.1 Taylor Approximation

Suppose we have a scalar valued velocity function v (*i.e.* speed) only dependent on time, *i.e.* $v = v(t)$. Its value at a later time $v(t + \Delta t)$ can be calculated through Taylor approximations [7] by

$$v(t+\Delta t)=v(t)+\frac{dv}{dt}\Delta t+\frac{d^2v}{dt^2}\frac{\Delta t^2}{2}+\frac{d^3v}{dt^3}\frac{\Delta t^3}{6}+\frac{d^4v}{dt^4}\frac{\Delta t^4}{24}+\ldots$$

Usually, the above is represented as:

$$v(t+\Delta t)=v(t)+\frac{dv}{dt}\Delta t+R \tag{2}$$

where the term R is the remainder (or the error) in ignoring higher order differentials in v with respect to time t. This error is the one source of error in CFD (one of many as discussed below) and is present regardless if the **Finite Difference Method** (FDM), the **Finite Volume Method** (FVM) or the **Finite Element Method** (FEM) is used. This is because all methods use finite differences (of one form or another) in the time marching algorithm [3,8] (*i.e.* when estimating the differential of velocity with respect to time). The FDM also uses finite differences in geometrical representations, but the FVM uses so-called *finite volumes*, while the FEM has *finite elements*.

Now, Eq. (2) can be rearranged into:

$$\frac{v(t+\Delta t)-v(t)}{\Delta t}=\frac{dv}{dt}+\frac{R}{\Delta t}=\frac{dv}{dt}-O(\Delta t)$$

where $O(\Delta t) = - R/\Delta t$ is the order of magnitude of error because of the Taylor series are being truncated after the first time derivative of the velocity. Thus, $O(\Delta t)$ represents the truncation error and its value can either be positive or negative. The hope is that $O(\Delta t)$ is very small for a given calculation. Thus, from the above equation, the estimation of $dv(t)/dt$ is now given by:

$$\frac{dv}{dt}=\frac{v(t+\Delta t)-v(t)}{\Delta t}+O(\Delta t)$$

Thus the differential $dv(t)/dt$ can be approximated with

$$\frac{dv}{dt}=\frac{v^{k+1}-v^{k}}{\Delta t}+O(\Delta t)\approx\frac{v^{k+1}-v^{k}}{\Delta t} \tag{3}$$

In the above equation, the time points $t + \Delta t$ and t have been replaced with time index k in the sense that $t = k\cdot\Delta t$ and $(k+1)\cdot\Delta t = k\cdot\Delta t + \Delta t = t + \Delta t$. This is a customary approach in CFD and should be familiar to the reader of this document. As will be clear shortly, the term Δt represents the time step used in the numerical simulation. The above scheme is called forward difference in time, but other implementations exists as well but are not treated, to maintain the simplicity of the current text. If it is desirable to estimate the differential $d\alpha/dy$ where $\alpha = \alpha\,(y)$ is for example a physical quantity (see Fig. 4.1, about the y-direction), then the same steps can be used that led to Eq. (3), giving the following:

$$\frac{d\alpha}{dy}=\frac{\alpha_{i+1}-\alpha_{i}}{\Delta y}+O(\Delta y)\approx\frac{\alpha_{i+1}-\alpha_{i}}{\Delta y} \tag{4}$$

where Δy represents the grid spacing.

4.2.2 A Very Simple CFD Example – Automatic Generation of Errors

To adequately demonstrate the sources of errors in CFD, we will use a very simple example of flow. Also, since the equation of motion Eq. (1) represents in fact Newton's second law, we will use the latter in its basic form, namely with:

$$m\frac{dv}{dt} = F \tag{5}$$

The term m is in this case the mass of a fluid particle (or *group particle*) and its value is fixed, meaning m = constant (see [9] in Section 4.2.2, about the fluid particle and the direct connection between Eqs. (1) and (5)). The term v is the (scalar valued) velocity (*i.e.* the speed) of the fluid particle and F is the sum of external forces applied to the fluid particle from its surroundings (here, we will ignore body forces like gravity). For simplicity, the velocity is assumed to be a scalar valued function of flow in x-direction (see Fig 4.1), only dependent on the time t (in reality, the speed v would also depend on spatial coordinates like the y–coordinate meaning $v = v(y,t)$, but we will ignore this for simplicity). Thus v represents the speed of the fluid particle. In this relation, it should be noted that Eq. (5) is only a differential equation, but for a real flow problem we would be in fact working with one or several coupled PDEs, generated from Eq. (1), as well as from other laws, like the continuity equation and the energy equation (if needed).

Consider the flow problem shown in Fig. 4.1. For this particular case, the problem consists of calculating the flow (*i.e.* the speed $v = v(t)$) between two stationary wall boundaries. We assume that Eq. (5) is valid, where F is given by

$$F = (d\alpha/dy)\cdot v(t) \tag{6}$$

The term α could for example represents some sort of viscous dissipation as well as pressure contribution that depends on location, meaning $\alpha = \alpha(y)$. Again, the above Eqs. (5) and (6) are just simple examples, which are not related to any real problem, but are good enough to adequately discuss the source of numerical errors. The point here is to demonstrate the fundamentals of CFD so its source of errors can be shown and discussed.

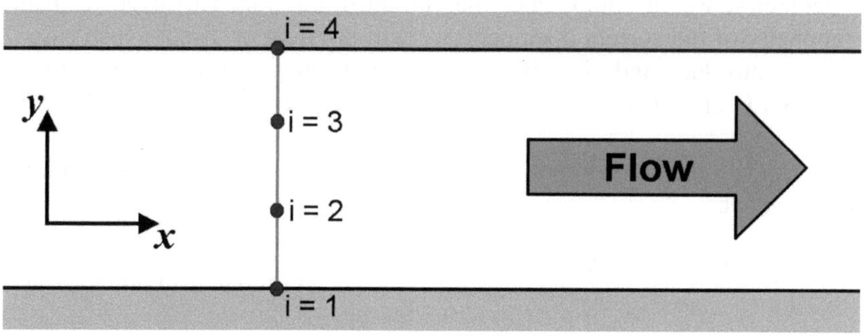

Fig. 4.1 Flow of a fluid between two stationary wall boundaries

A computer cannot understand the problem as presented with Eq. (5) and Fig. 4.1. Therefore, both must be converted to something that makes that possible. To do this Fig. 4.1 is represented with grid points, marked as i = 1, i = 2, i = 3 and i = 4. At each of these points, the speed v has to be determined. These are represented with v_1, v_2, v_3 and v_4, respectively (also represented as v_i in which i = 1,2,3,4). Not only do we want to calculate the speed at these four grid points, but also its time evolution. As explained with Eq. (3), the time is represented with the time index k, thus in fundamental terms we want to calculate v_i^k where i = 1,2,3,4 and k = 0,1,2,3,…N, in which the total simulation time is $t_{tot} = N \cdot \Delta t$. As explained before, the term Δt is the time step used in the numerical simulation and is a very important factor in terms of numerical stability.

Assuming no slippage at the wall boundary, we can immediately put $v_1^k = 0$ and $v_4^k = 0$ for all time steps k (a so-called Dirichlet boundary condition is being implemented). But at the grid points 2 and 3, the speed has to be calculated according to Eq. (5). To do so, this equation must be converted to algebraic equation and this is done by Eqs. (3) and (4), giving:

$$m\left[\frac{v_i^{k+1} - v_i^k}{\Delta t}\right] = \left[\frac{\alpha_{i+1} - \alpha_i}{\Delta y}\right] \cdot v_i^{k+1} \quad \text{...for i = 2 and 3} \tag{7}$$

Notice the obvious similarities between Eqs. (5) and (7). After some rearrangements, the above is written as:

$$\left(1 - \left[\frac{\alpha_{i+1} - \alpha_i}{m}\right] \cdot \frac{\Delta t}{\Delta y}\right) v_i^{k+1} = v_i^k \qquad \text{...for i = 2 and 3} \tag{8}$$

Because of the use of Eq. (4), the above CFD scheme is the Finite Difference Method (FDM). Although the main focus is here on the FDM, the concept of CFD error is the same for both FEM and the FVM. Discussing the sources of error by the FDM simplifies the current text considerably without losing objective. Thus for all methods FDM, FEM or FVM, the term Δy represents the distance between grid points.

Generally speaking, the spacing between grid points Δy is varied through the flow geometry, which complicates the current discussion. However, to maintain the simplicity of the current document, we will not treat or discuss such case. The point is to introduce and give overview of CFD error, and not get into the finer points of CFD error (the reader is rather redirected to the reference given in the back).

From Eq. (8), then at grid point 2 (*i.e.* replacing the grid number i with 2 in the above), we have

$$\left(1 - \left[\frac{\alpha_3 - \alpha_2}{m}\right] \cdot \frac{\Delta t}{\Delta y}\right) v_2^{k+1} = v_2^k \qquad \text{(valid at grid point i = 2)}$$

and likewise, for grid point 3, we have

$$\left(1-\left[\frac{\alpha_4-\alpha_3}{m}\right]\cdot\frac{\Delta t}{\Delta y}\right)v_3^{k+1}=v_3^{k} \qquad \text{(valid at grid point i = 3)}$$

Generally for CFD, equations like the two above are grouped together into a single array equation, also known as ***system of algebraic equations*** (in fact, for a real CFD problem, there would be several such equations):

$$\begin{pmatrix} 1-\left[\frac{\alpha_3-\alpha_2}{m}\right]\cdot\frac{\Delta t}{\Delta y} & 0 \\ 0 & 1-\left[\frac{\alpha_4-\alpha_3}{m}\right]\cdot\frac{\Delta t}{\Delta y} \end{pmatrix}\begin{pmatrix} v_2 \\ v_3 \end{pmatrix}^{k+1}=\begin{pmatrix} v_2 \\ v_3 \end{pmatrix}^{k} \tag{9}$$

The above equation can be written as

$$\mathbf{A}\cdot\mathbf{v}^{k+1}=\mathbf{b}(\mathbf{v}^{k}) \tag{10}$$

For the simple case presented here, then $\mathbf{b}(\mathbf{v}^k) = \mathbf{v}^k$ but this is generally not so. That is, the vector function **b** will depend on $\mathbf{v}^k$, as well as on other quantities, physical and numerical, just as the array **A** does (see for example Section 7.4 in [9] about a realistic presentation of Eq.(10) that applies for the coaxial cylinders viscometer). To solve the system in Eq. (10), methods like the Cramers's rule, Gaussian elimination, or Jacobi and Gauss-Seidel point-iterative methods can be used [3]. The type of algorithm that exists to solve Eq. (10) is a large theme and is beyond the scope of this text. For whatever method a CFD software uses, the particular solution algorithm will be presented here as $\mathbf{v}^{k+1}= \mathbf{A}^{-1}\cdot\mathbf{b}$, in which theoretically $\mathbf{A}^{-1} = \mathrm{adj}(\mathbf{A})/\det(\mathbf{A})$ [10].

The value of **v** at the time step k, namely $\mathbf{v}^k$, is always known, starting with $\mathbf{v}^{k=0}$ that defines the initial condition. From this information, we want to calculate the unknown speed at the time step k+1, giving $\mathbf{v}^{k+1}$, by Eq. (10). <u>This last mentioned equation describes the CFD in the most fundamental terms</u>. Regardless if the used method is FDM, which was actually used above for Eq. (9), FVM or FEM, all of these methods consist of converting equation like Eq. (5) into a system of algebraic equations $\mathbf{A}\cdot\mathbf{v}^{k+1}= \mathbf{b}(\mathbf{v}^k)$ like shown with Eq. (10). For the same geometrical problem and the same setup of grid points, the numerical value of each component of **A** and **b** will differ somewhat, as well as the location of non-zeros in the array **A**. But regardless of that, all methods (FDM, FEM, FVM and so forth) should give the same or very similar results. The time scheme shown in Eq. (9) is called *implicit scheme* (regardless if FDM, FVM or FEM is used). There exist another method called the *explicit scheme* and is much simpler than the implicit method (in which Eq. (10) is not produced). Because the explicit scheme incorporates much larger numerical instability in CFD calculations, it is generally not used when calculating the flow of non-Newtonian fluids, like of the cement based material, and thus is not relevant here.

4.3 Numerical Errors (E1)

With Eqs. (9) and (10) and the steps that lead to these equations, errors in CFD can now be addressed.

4.3.1 Discretization Error

Although Eq. (9) represents Eq. (5), these two equations are NOT the same and their differences constitute a so-called ***truncation error***. That is, the difference between Eqs. (5) and (9) rests in the use of Eqs. (3) and (4), where the terms $O(\Delta t)$ and $O(\Delta y)$ have been ignored[1]. Because of this difference, the differential Eq. (5) will produce a different result, relative to the result produced by Eq. (9).

If the (unknown) result produced by Eq. (5) is represented with v, and the result produced by Eq. (9) with v_i^k then the following applies:

$$v = v_i^k + e_i^k \tag{11}$$

The term e_i^k is then the ***discretization error*** and originates directly from the *truncation error* mentioned just above (*i.e.* from $O(\Delta t)$ and $O(\Delta y)$). In general, as number of time and spatial grid points goes to infinity ($\Delta t \rightarrow 0$ and $\Delta y \rightarrow 0$) the discretization error will converge to zero ($e_i^k \rightarrow 0$). Increasing the number of spatial grid points is called ***grid refinement***.

Grid refinement is the main tool at the disposal of the CFD user for the improvement of the accuracy of a simulation. The user would typically perform a simulation on a coarse mesh first to get an impression of the overall feature of the solution. Subsequently the grid is refined in stages until no (significant) differences of results occur between successive grid refinement stages. Results are then called ***grid independent*** [3].

To demonstrate grid refinement and grid independence, the CFD software VVPF 2.0 is used (freely available at www.vvpf.net). In this case it is simulating the flow of a Bingham material inside a coaxial cylinders viscometer [11]. For this case, the system of algebraic equations is much more complicated than presented with Eq. (9). These equations can be seen in [9] (Section 7.4) and can be downloaded from http://www.diva-portal.org/. Because of the geometry in question, cylindrical coordinate system is used.

Fig. 4.2. gives a concise visual summary of the flow problem to be solved for the above mentioned viscometer. The inner cylinder is stationary and registers torque T [Nm], while the outer cylinder rotates at predetermined rotational frequency f_o [rps]. On the upper–right illustration, the flow is shown in a more scientific manner, as a plot of the velocity (or rather speed in θ–direction) as a function of radius r.

---

[1] Depending on the particular method used, a higher order of accuracy can be attained in FDM. For example, second order of accuracy in time can be produced, then represented as $O(\Delta t^2)$. To maintain simplicity of the current text, such case is not treated here.

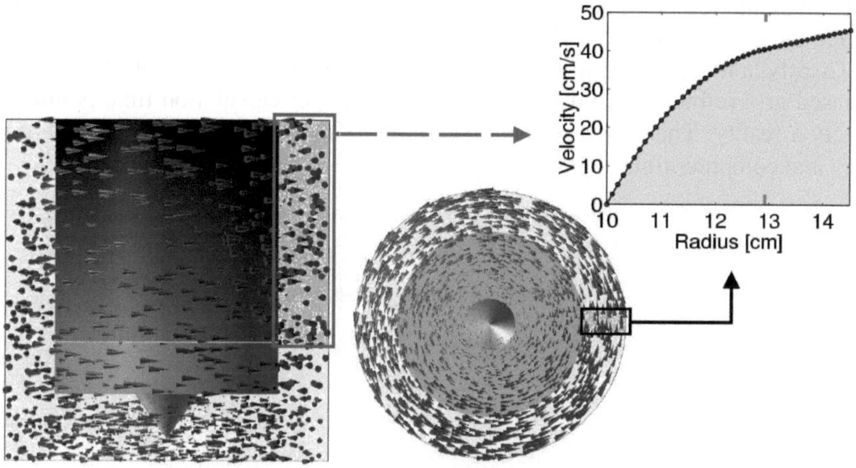

Fig. 4.2 A concise visual summary of the flow problem for the coaxial cylinders viscometer

In Fig. 4.3a it is demonstrated how changed grid refinement alters the calculated torque T applied on the inner cylinder (see Eq. (12) in [11] about how torque is calculated). For this case, the yield stress τ_0 is set to 800 Pa, while the plastic viscosity μ is equal to 20 Pa·s. The rotational frequency f_o of the outer cylinder is set 0.5 rps.

With more and more grid points used in the calculations (*i.e.* smaller grid spacing Δr), an equilibrium value in torque T is attained, very close to the analytical correct value, shown with the horizontal blue line. Among others, because the corresponding PDE is not the same as the system of algebraic equations, a complete match between the theoretical correct torque value and the computed torque value is never attained.

The peculiar oscillation in computed torque as a function of grid spacing Δr shown in Fig. 4.3a is most likely occurring when grid points are not at or close to the theoretical plug location. This is more clearly shown in Fig 4.3c. This last mentioned illustration shows the apparent viscosity η as a function of radius r. The blue vertical line shows the theoretical correct location of plug, calculated as explained in [11]. In Fig. 4.3b is the velocity shown for the different cases of grid spacing Δr, while the shear rate is shown in Fig. 4.3d. Finally, the shear stress is shown in Fig. 4.3e. In Figs. 4.3f to 4.3i, the correct analytical result is shown for comparison, calculated as explained in [11] (see also for example Sections 3.3.4 and 3.5.2 in [9]). Also shown in these last mentioned illustrations is the numerical case with grid spacing of Δr = 2 mm. The two results are more or less identical, with however a slight deviation apparent in Fig. 4.3g.

In Figs. 4.3b to 4.3e, every tenth result is only shown (to clearly present the results), while in Fig. 4.3a, all results of the experiment are shown. The number of simulations done in this experiment is 178, going from Δr = 0.2 mm up to Δr = 17.9 mm in the steps of 0.1 mm. The whole experiment was automatically generated through two interlinked Python (2.6) scripts, in which one of them called the (Fortran) binary 178 times.

Clearly, a too coarse grid spacing Δr gives a bad simulation result. But with increased grid refinement and better accuracy, a longer calculation time is unfortunately a reality. Thus, the user has to carefully choose the balance between accuracy and computer time.

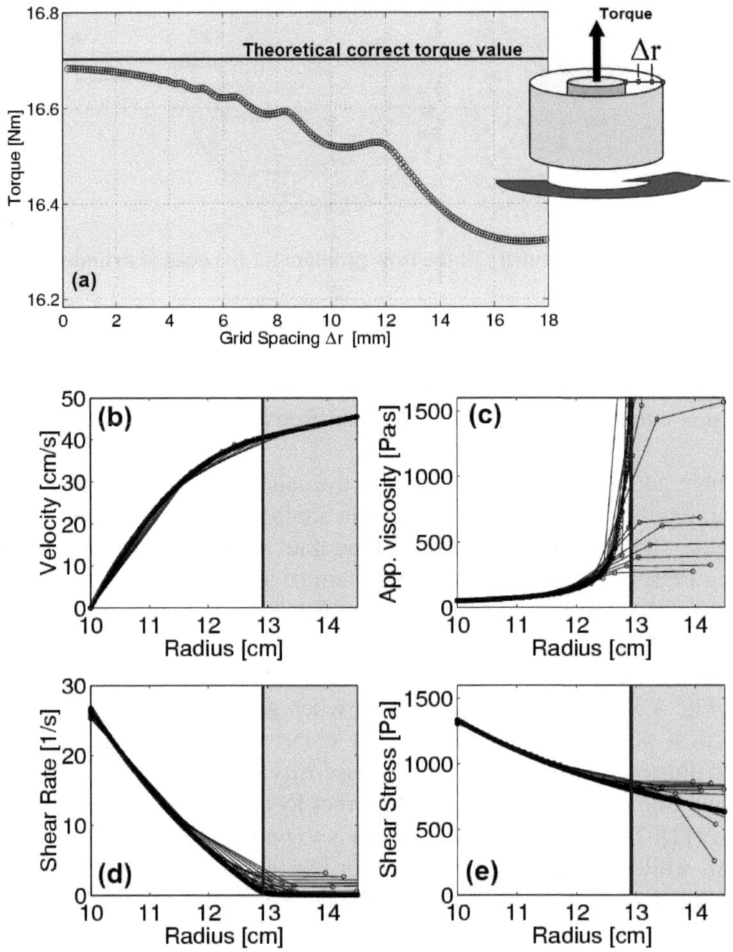

Fig. 4.3 The effect of increased grid refinement on calculated flow. The gray area in illustrations (b) to (i) represents the true plug (see Fig. 4.2. for concise visual summary of the system).

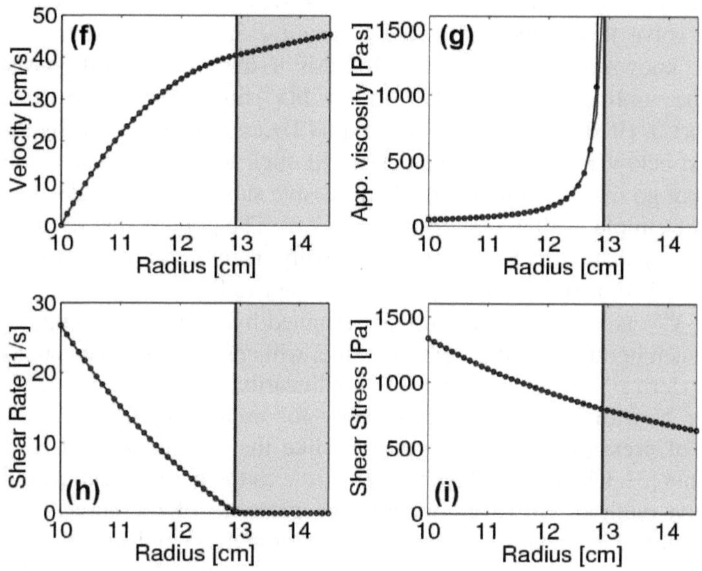

Fig. 4.3 (*continued*)

4.3.2 Iterative Convergence Errors

Numerical solution of a flow problem requires an iterative process. Depending on the flow problem at hand, the iterative process can be quite different from one case to another. For example for steady state problem where pressure and velocity are coupled, some sort of iteration is required to correct both velocity and pressure (see for example the SIMPLE and SIMPLER algorithms in [3]). This applies also for time dependent calculations, in which SIMPLE, SIMPLER or PISO (and so forth) are used. These techniques are too comprehensive to discuss here. Instead we will focus on a method which does not involve pressure, or pressure correction. This will not diminish the text, as in fundamental terms the discussion made here will also apply for the above mentioned techniques.

The example given here, is relative to the problem presented with Eq. (9) (or equally, with Eq. (10)), which are (in basic terms) very typical for flow problems of viscoplastic fluid (*i.e.* for flow of cement based material).

Calculating the flow for the next time step k+1 in Eq. (10) is straightforward primarily only because all elements in the array **A** are known. Thus, the next velocity step is calculated as $\mathbf{v}^{k+1} = \mathbf{A}^{-1} \cdot \mathbf{b}(\mathbf{v}^k)$. However, when calculating the flow of a viscoplastic fluid, things become considerably more complicated. The main reason for this is because the array **A** depends on the primary unknown, namely on $\mathbf{v}^{k+1}$. This means that the problem to be solved is a nonlinear system of algebraic equations, given by:

$$\mathbf{A}(\mathbf{v}^{k+1})\cdot\mathbf{v}^{k+1} = \mathbf{b}(\mathbf{v}^k) \qquad (12)$$

In fundamental terms, since $\mathbf{v}^{k+1}$ is an unknown, the term $\mathbf{A}^{-1}$ cannot be calculated. To solve this, a method called *Successive Substitution Method* [8] can be used, also known as Picard iteration. In basic terms, this method consists of converting the nonlinear system $\mathbf{A}(\mathbf{v}^{k+1})\cdot\mathbf{v}^{k+1}= \mathbf{b}(\mathbf{v}^k)$ into a sequence of linear ones $\mathbf{A}\cdot\mathbf{v}^{k+1}= \mathbf{b}(\mathbf{v}^k)$. [It should be noted that Eq. (12) can also be solved by the methods mentioned below Eq. (10), but here we will stick to the successive substitution.] We will not go into the fine points of successive substitution, but detailed information is for example available in [9] (Section 7.8). The point is that for a given calculation, namely from k to k+1, there are actually many iterations involved in which $\mathbf{v}^k$ is frozen. The next time step, namely from k+1 to k+2, is not started until the accuracy of $\mathbf{v}^{k+1}$ is sufficient, which is determined by a specific tolerance value. Increasing such iteration during each time step, will cost prolonged calculation time.

For modern FVM, for example, the nonlinearity shown in Eq. (12) is also produced for Newtonian flows (*i.e.* not only for non-Newtonian flows). With the presence of pressure gradient, algorithms like the time-dependent SIMPLE (see for example [3], Chapter 8) plays the same role as the above mentioned successive substitution method. However, without the presence of pressure gradient in the flow direction (like what often applies for viscometers) and thus the lack of need for SIMPLE, the successive substitution method can be used instead. This is done for the software VVPF (www.vvpf.net). Thus roughly speaking, the discussion below about the successive substitution method could also be considered to apply for the time-dependent SIMPLE and similar algorithms.

An increased number of successive substitutions for a given time step will increase the accuracy of the overall simulation. But the cost will be longer simulation time. Again, as mentioned in Section 4.2.1, the user has to carefully choose the correct balance between accuracy and prolonged computer time. To demonstrate the above point, the software VVPF 2.0 is used, in which its result is shown in Fig. 4.4. This figure shows the shear rate profile for time-dependent flow simulation inside a coaxial cylinders viscometer, using the so-called PFI material model [18]. Inner radius is 8.5 cm and is stationary, while the outer radius is 10.1 cm and rotates at predetermined angular velocities as explained in [18]. The zero shear rate represents plug (*i.e.* rigid body rotation).

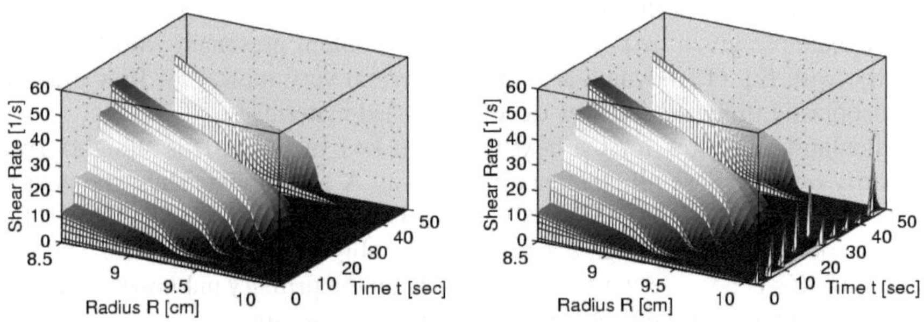

Fig. 4.4 The influence of the number of successive substitutions on the simulation quality (for comparison, time independent shear rate for Bingham fluid is shown in Fig. 4.3h, for example)

In the left illustration of Fig. 4.4, the maximum allowed number of successive substitutions is 1000 (a tolerance value determines the actual number of successive substitutions), while on the right illustration this number is 1. That is, for the latter case, no successive substitution is actually allowed, generating unwanted oscillations in shear rate (or a iterative convergence error) in the plugged zone. However, the simulation speed is about 5 times faster for this latter case. In terms of calculated torque, the influence of this error is minimal because it is calculated at the inner cylinder. But if the torque would have been calculated at the outer cylinder, the effect of this error would be catastrophic. So it depends on what is being simulated and where the numerical error is occurring in the flow, if the numerical error is fatal or not for a particular CFD problem.

4.3.3 Round Off Errors

As previously mentioned, an equation like Eq. (10) represents the core CFD system that is solved in a computer. That is, a single or several such equations represent the center of a CFD software, regardless if FDM, FVM or the FEM is used. The size of the array **A** for a real CFD problem can easily be, say 10000x10000, or even much larger (and not 2x2 like shown with Eq. (9)).

The round-off error exists because the computer cannot give answers to an infinite number of decimal places. Every calculation that is made in a CFD software is carried out to a finite number of significant figures. This introduces a round-off error at each step of the computation [12.13]. However, the round off error in a well written CFD software, is reduced (however, NOT avoided!) by avoiding subtraction (*i.e.* –) of almost equal sized large numbers and by avoiding the addition (*i.e.* +) of numbers of with very large difference in magnitude [3].

From the above text, it should be clear that the computational solution of the system of algebraic equations Eq. (10) (*i.e.* Eq. (9)) is not $v_{\mathrm{i}}^{\mathrm{k}}$ (as it ideally should be) but rather $\mathrm{V}_{\mathrm{i}}^{\mathrm{k}}$ (which represents the value that the computed provides). Thus, we have the following relationship:

$$v_{\mathrm{i}}^{\mathrm{k}} = \mathrm{V}_{\mathrm{i}}^{\mathrm{k}} + \xi_{\mathrm{i}}^{\mathrm{k}} \qquad (13)$$

where the term $\xi_{\mathrm{i}}^{\mathrm{k}}$ represents the round-off error and can vary from different grid points i, and at different time steps k. Commonly in CFD literature, the solution of Eq. (5), namely v, is called the "***correct solution***". The ideal solution of Eq. (10) (or Eq. (9)), namely $v_{\mathrm{i}}^{\mathrm{k}}$, is commonly referred to as the "***exact solution***". As indicated above, the term $\mathrm{V}_{\mathrm{i}}^{\mathrm{k}}$ then represents the "***computed solution***", and is produced by the computer. These designations do however depend on what reference is used.

When the computed solution $\mathrm{V}_{\mathrm{i}}^{\mathrm{k}}$ approaches the exact solution $v_{\mathrm{i}}^{\mathrm{k}}$ for the given Δt and Δy stability is achieved in the numerical calculations [8,13]. More precisely, stability is the tendency for any spontaneous numerical perturbations $\xi_{\mathrm{i}}^{\mathrm{k}}$ in the computed solution $\mathrm{V}_{\mathrm{i}}^{\mathrm{k}} = v_{\mathrm{i}}^{\mathrm{k}} - \xi_{\mathrm{i}}^{\mathrm{k}}$ to decay. This perturbation consists mainly of the round-off error [12] and is not to be confused with the discretization error $e_{\mathrm{i}}^{\mathrm{k}}$, which is due to incorrect logic of the discretization method used, which are slightly different between FDM, FVM and FEM. In terms of accuracy (at least on simple grid systems), one method does not so much precedes the other. Which of the three basic methods are best, depends on geometry of the boundary condition, material model and on other factors. Comparison in accuracy between FDM and FEM is available in [13].

To summarize the above text: During the calculation of CFD, we want to gain *correct solution* v, (as produced by Eq. (5)), but because of the **discretization error** e_i^k the system of algebraic equations Eq. (10) (or equally, Eq. (9)), can theoretically only provide us with the *exact solution* v_i^k (see Eq. (11)). Furthermore because of the **round off error** ξ_i^k, the computer (that uses Eq. (10)), will only provide us with the *computed solution* V_i^k (see Eq. (13)).

Combining Eqs. (11) and (13) gives the relationship between the correct solution v (*i.e.* the solution we actually want) and the output of the computer, namely V_i^k (*i.e.* the solution we are actually attaining):

$$v = V_i^k + e_i^k + \xi_i^k \tag{14}$$

4.4 Coding Errors (E2)

With ever increasing complexities in material modeling, as well as higher demand of efficiency (programming speed) from programmers by their employee (*i.e.* by the particular CFD software company), programming mistakes (also known as *coding error*) can always be present in a CFD software. A successful comparison of known flow to simulated flow, does not guarantee fault free CFD code. This is because the code error might not emerge until a very specific type of flow, or type of material model is being used. For example, benchmarking a code by using Newtonian fluid flow might give an indication of correct code. However, going into viscoplastic flow in which shear rate is for example incorrectly calculated will render a false result. Continuing from Eq. (14), the relationship between the correct solution v (*i.e.* the solution we actually want) and the output of the computer, namely V_i^k (*i.e.* the solution we are actually attaining) is given by:

$$v = V_i^k + e_i^k + \xi_i^k + C_i^k \tag{15}$$

where C_i^k represents error generated by code error present in the CFD software. It should be clear that coding error can exist in any type of simulation packages, regardless of CFD, DEM or other types of fundamental scheme.

4.5 User Error (E3)

By the ever increasing popularity and availability of CFD softwares and the necessary hardware to run them, more and more unskilled personals will start to use such tools. The term "unskilled", basically means two things. Firstly, the lack of fundamental understanding of the mathematics of CFD and thus sources of errors (in which this document is aimed to help with). Secondly, insufficient training on a particular CFD software, leading to its improper or wrong use. Continuing from Eq. (15), the relationship between the correct solution v (*i.e.* the solution we actually want) and the output of the computer, namely V_i^k (*i.e.* the solution we are actually attaining) is given by:

$$v = V_i^k + e_i^k + \xi_i^k + C_i^k + U_i^k \tag{16}$$

where U_i^k represents error generated by insufficient training and/or lack of fundamental understanding of the mathematics of CFD (and thus sources of errors). For any given flow simulation, the aim is to maintain $v \approx V_i^k$. It should be clear that the sum $e_i^k + \xi_i^k + C_i^k + U_i^k$ will never be zero and this is neither the objective. The objective is however, to keep this sum as low as possible.

4.6 Error from Input Uncertainties (U1)

General input uncertainties constitute issues like the difference between the geometry used in the CFD calculation. For example, the assumed dimensions of the formwork setup, including the location of the rebars, during the pre-processing can easily differ by many millimeters to the actual formwork and rebar location used at a particular building site. Human building tolerances during the construction of mold or formwork will lead to differences between calculated flow from CFD and the actual observed flow. The significance of this error will depend on the complexities of the geometry (*i.e.* the complexities of the formwork or mold, rebar location and its number) in question.

Boundary conditions represent another input uncertainty that is of much greater relevance in CFD casting predictions. For cement based material, it is very hard to postulate with full certainty how much slippage is actually occurring at solid boundaries for a given case. Assuming perfect slippage at smooth surfaces can be questionable in some cases due to the varying degree of stickiness behavior of the cement based material. On the flip side of the problem, the presence of coarse aggregates in fresh concrete can easily generate slip layer, producing slip velocity at the wall boundary. That is a slip layer is present, where the velocity evolves from zero at the wall to the apparent slip velocity at the boundary of the slip layer. Still, for such cases, a full slippage cannot be assumed and some sort of friction parameter has to be determined. Thus, the input parameters for the simulation are not only the physical material properties of the cement based material, but also a friction parameter that depends on the properties of the wall boundary as well as if wall layer is forming (other factors will also enter in this problem).

Unfortunately, an outcome from a CFD can be significantly changed by changing the friction parameters that define the degree of slippage. Ironically, the friction parameter can be artificially fitted, so a similar behavior is observed between actual flow and the CFD flow. The above mentioned error is perhaps the most problematic one in CFD for fresh concrete, and more research related to this issue is of paramount importance.

4.7 Physical Model Uncertainty (U2)

4.7.1 Choosing the Correct Material Model

It is generally agreed that the Bingham model represents the flow properties of fresh concrete, at least as a good first approximation. More precisely, it is commonly agreed that the fresh concrete can with good accuracy be considered as Bingham fluid [14-17]. The Bingham fluid represents an important class of viscoplastic material. It possesses a yield stress that must be exceeded before significant deformation can occur. Such materials typically sustain an applied stress at rest. Generally speaking, such materials are referred to as viscoplastic material.

In some cases (perhaps many), the Herschel-Bulkley model (another type of viscoplastic material) can be more suitable to describe the material behavior of the fresh concrete. For example, the use of the Herschel-Bulkley model might show an important flow condition (say at corners?) that might otherwise not be picked up when using the Bingham model. Deciding or determining what model is the best suitable (say, for a specific concrete batch) constitutes the one part of the physical model uncertainty. But this type of uncertainty is easily resolved by repeating the whole simulation with two different material models.

4.7.2 Implementation of Yield Stress

As the fresh concrete is a "yield stress fluid" (*i.e.* viscoplastic material), the final input uncertainty discussed here, will be about the numerical implementation of such material model, and the problems associated with it. Although the emphasis will be on the Bingham model, the discussion applies equally to other types of viscoplastic models like the Herschel-Bulkley, or the very complicated PFI-model [18].

4.7.2.1 A Theoretically Correct Bingham Presentation

The extra stress tensor for a Bingham fluid, employing a von Mises yield criterion for flow, was originally formulated by Oldroyd [19,20], however using syntax for example given by Malvern [6] it can be represented with:

$$\mathbf{T} = \left(\mu + \frac{\tau_0}{\sqrt{\mathrm{II_d}}} \right) \mathbf{d} \qquad \text{for } \mathrm{II_S} \geq \tau_0^2 \tag{17}$$

$$\mathbf{d} = 0 \qquad \text{for } \mathrm{II_S} < \tau_0^2 \tag{18}$$

where $\mathbf{d} = \nabla\mathbf{v} + (\nabla\mathbf{v})^{\mathrm{T}}$ is the rate-of-strain tensor and **v** is the three directional and four dimensional velocity (the time being one dimension). The term μ is the plastic viscosity and τ_0 is the yield stress. Here, the (total) stress tensor $\boldsymbol{\sigma}$ has been decomposed into an isotropic pressure term p and an extra stress contribution **T**, commonly known as the extra stress tensor [21]. That is, $\boldsymbol{\sigma} = -p\,\mathbf{I} + \mathbf{T}$, where **I** is the unit dyadic. The term $\mathrm{II_S}$ is the second invariant of the deviator stress tensor, given by[2] [6]

$$\mathrm{II_S} = -[\mathrm{tr}(\mathbf{S})\,\mathrm{tr}(\mathbf{S}) - \mathbf{S}:\mathbf{S}]/2 \tag{19}$$

where **S** is the deviator stress tensor, defined as [5,6]

$$\mathbf{S} = \boldsymbol{\sigma} - \frac{\mathrm{tr}(\boldsymbol{\sigma})}{3}\mathbf{I} \tag{20}$$

With Eq. (20), then tr(**S**) = 0 and thus $\mathbf{II_S} = \mathbf{S}:\mathbf{S}/2$. Also, for incompressible material, then **S** = **T** and therefore $\mathrm{II_S} = \mathbf{S}:\mathbf{S}/2 = \mathbf{T}:\mathbf{T}/2$ [11]. The term $\mathbf{II_d}$ is the second invariant of the rate-of-strain tensor, given by [6]

$$\mathrm{II_d} = -[\mathrm{tr}(\mathbf{d})\,\mathrm{tr}(\mathbf{d}) - \mathbf{d}:\mathbf{d}]/2 \tag{21}$$

The above equation represents the rate of shear intensity for three dimensional flow, sometimes/usually referred to as the shear rate. That is, $\mathrm{II_d} = \dot{\gamma}^2$, where $\dot{\gamma}$ represents the shear rate, or the deformation intensity of the flow [8,22]. According to Eqs. (17) and (18), the yield surface is located by the condition $\mathrm{II_S} = \tau_0^2$. In the region where $\mathrm{II_S} < \tau_0^2$, the material behaves as a rigid solid (*i.e.* plug), while in the region with $\mathrm{II_S} > \tau_0^2$, the material flows with a shear viscosity of $\eta = \mu + \tau_0/\sqrt{\mathrm{II_d}}$. With the shear viscosity η, then Eq. (17) can be rewritten to T = $\eta(\mathrm{II_d})$ **d**, which constitute a so-called Generalized Newtonian Model (GNM) [23].

4.7.2.2 Viscoplastic Implementation for CFD

The Regularization Approach

Because of the non-linearity in the governing equation and because of the inherent discontinuity in the constitutive equation, a computer simulation of Bingham flow is difficult. As the yield surface is approached, the presence of $\mathbf{II_d}$ in the

---

[2] Here, it is assumed that the characteristic equation is $\lambda^3 - \mathrm{I}\,\lambda^2 - \mathrm{II}\,\lambda - \mathrm{III} = 0$, where I, II and III are the first, second and third invariants, respectively [6]. In [5,11], the characteristic equation is $\lambda^3 - \mathrm{I}\,\lambda^2 + \mathrm{II}\,\lambda - \mathrm{III} = 0$, and thus the second invariant is given by $\mathrm{II_s}$ = [tr(**S**)tr(**S**) – **S**:**S**]/2. For that case, the shear stress (in the second power) is calculated as –$\mathrm{II_s}$. (i.e. not as $\mathrm{II_s}$). Regardless of what presentation is used, the same (second power) shear stress is attained –[tr(**S**)tr(**S**) – **S**:**S**]/2.

denominator of Eq. (17) makes the apparent viscosity unbounded [11]. Furthermore, while simulating the velocity field **v**, the location of the yield surface is unknown prior to calculation. To overcome these difficulties, a regularized version of the Bingham model has been proposed by Bercovier and Engelman [24]. It consists of adding a small regularization parameter δ in the denominator of Eq. (1). With this, the discontinuous model Eqs. (17) and (18) are transformed to the continuous model

$$\mathbf{T} = \left(\mu + \frac{\tau_0}{\sqrt{\mathrm{II}_{\mathbf{d}}} + \delta} \right) \mathbf{d} \tag{22}$$

Bercovier and Engelman used Eq. (22) to solve Bingham flow in a closed square cavity subject to a body force [24]. This equation has also been used by Taylor and Wilson to simulate conduit flow of an incompressible Bingham fluid [25], while Wallevik used it to simulate Bingham flow inside a coaxial cylinders viscometer [11]. Furthermore, Burgos *et al.* used the regularization parameter δ in this manner to simulate antiplane shear flow of a Herschel-Bulkley fluid [26]. As reported in [11], a lower value of δ gives a better imitation of true Bingham behavior.

The Method of Papanastasiou

Papanastasiou [27] introduced an alternative regularization approach, in which Eqs. (17) and (18) are transformed into

$$\mathbf{T} = \left(\mu + \frac{\tau_0 \left[1 - \exp(-m\sqrt{\mathrm{II}_{\mathbf{d}}}) \right]}{\sqrt{\mathrm{II}_{\mathbf{d}}}} \right) \mathbf{d} \tag{23}$$

In this case, the term m is now the regularization parameter for the Papanastasiou representation (*i.e.* not mass, *cf.* Eq. (5)), in which a higher value gives a better imitation of the Bingham fluid [26].

Direct Computational Implementation

A more direct computational implementation of Eqs. (1) and (2) is possible, which consisting of generating plateau (*i.e.* a max value) for the shear viscosity η. That is, the extra stress tensor is given by

$$\mathbf{T} = \eta(\mathrm{II}_{\mathbf{d}})\, \mathbf{d} \tag{24}$$

in which

$$\eta = \begin{cases} \mu + \dfrac{\tau_0}{\sqrt{\mathrm{II}_{\mathbf{d}}}} & \text{for} \quad \eta \le \eta_{\max} \\ \eta_{\max} & \text{for} \quad \eta > \eta_{\max} \end{cases} \tag{25}$$

A higher value of η_{max} gives a better imitation of the Bingham fluid. The above presentation is equivalent to the presentation given by O'Donovan and Tanner [28], in which a critical shear rate $\dot{\gamma}_c$ is rather used to control viscoplastic condition. The relationship between η_{max} and $\dot{\gamma}_c$ for the Bingham material is as simple as $\eta_{max} = \mu + \tau_0/\dot{\gamma}_c$. With this equation, then in Fig. 4.5, the corresponding critical shear rate value is calculated from the given value of η_{max}.

For all cases, *i.e.* either with Eq. (22), (23) or (24), the same material model is now used in simulating both the yielded region and the unyielded region [24-26].

4.7.2.3 Comparison of Different Viscoplastic Implementations

In Fig. 4.5, the apparent viscosity η (*i.e.* shear viscosity) is shown as a function of shear rate $\dot{\gamma}$ for the regularization approach Eq. (22), the method of Papanastasiou Eq. (23) and for the direct computational implementation Eqs. (24) to (25). As mentioned below Eq. (21), then $\mathrm{II}_{\mathbf{d}} = \dot{\gamma}^2$.

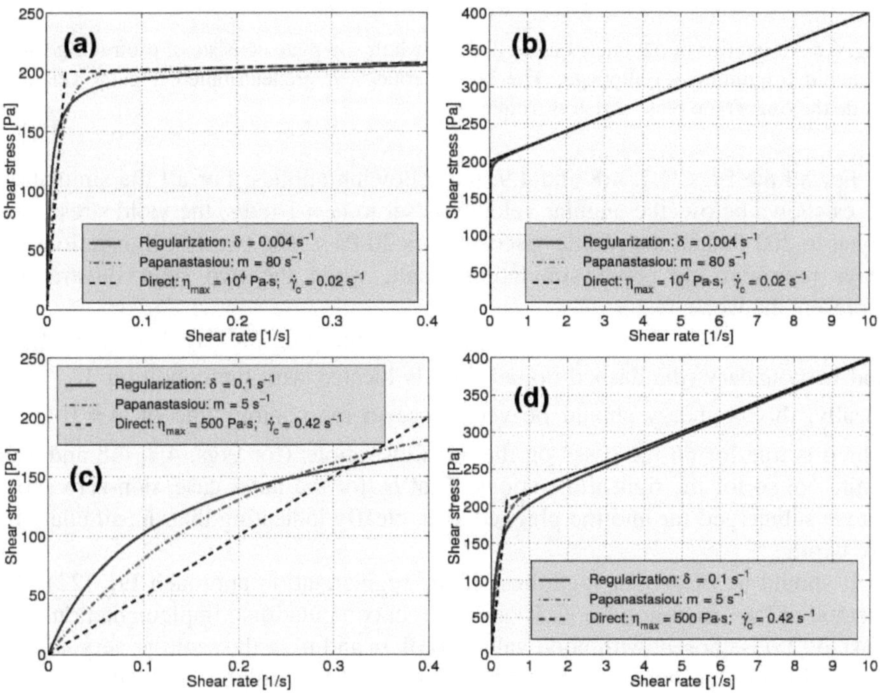

Fig. 4.5 Apparent viscosity (*i.e.* shear viscosity) as a function of shear rate for the different cases of the regularization approach, the method of Papanastasiou and the direct computational implementation for $\tau_0 = 200$ Pa and $\mu = 20$ Pa·s

Figs. 4.5a and 4.5b show better (*i.e.* more accurate) implementations of the Bingham fluid, while Figs. 4.5c and 4.5d show worse implementations. This is done by choosing appropriate values for δ, m and η_{max} shown in the figure. A better implementation will generate longer computational time, due to larger instabilities that will follow such steps.

To give an example for both good and bad quality simulation results, the flow inside a special type of parallel plate rheometer is analyzed [9,29]. The device is shown in Fig. 4.6. There, the top plate is stationary and registers torque, while the bucket rotates at predetermined angular velocity. The extremities of the vanes in the bucked and of the whole top plate defines the solid boundary (for further information about the simulation setup and boundary conditions, see [9,29]).

Fig. 4.6 All interior of the bucket as well as the whole top plate consists of protruding vanes to avoid (or minimize) slippage. The bucket rotates at predetermined angular velocity, while the (inner) top plate registers torque T.

For all the Figs. 4.7, 4.8 and 4.9, the following applies: For all the simulation cases shown below, the angular velocity is set to ω = 1 rad/s, the yield stress τ_0 is equal to 200 Pa and the plastic viscosity μ is 20 Pa·s. The two left illustrations always represent the better simulation result, while the two right illustrations represent the worse result.

The gray area in Figs. 4.7 to 4.9, represents the plug (*i.e.* rigid body rotation) and its boundary (the dashed dotted line) is located with the condition $\mathrm{II}_S = \tau_0^2$. Ideally, this boundary should be very close to the contour line of $\dot{\gamma} = 0.1\ \mathrm{s}^{-1}$, which is true for all the cases on the left illustrations (for Figs. 4.7, 4.8 and 4.9) while not so for the right illustrations. That is, for the latter case, non-zero shear rate is submerged far into the plugged area, clearly indicating that the simulations are wrong.

It should be clear that regardless if the regularization approach Eq. (22), the method of Papanastasiou Eq. (23) or the direct computational implementation Eqs. (24) to (25) is chosen, with good values for δ, m and η_{max}, the same or very similar simulation results should be attained. This is clearly accomplished for the left illustrations of Figs. 4.7 to 4.9. For these cases, the same calculated torque is gained (applied on the top plate), namely T = 4.99 Nm, while not so for the right illustrations. For the latter case, then $T = 4.74$ Nm is attained for δ = 0.1, $T = 4.86$ Nm for $m = 0.1$ and $T = 4.85$ Nm for $\eta_{max} = 500$ Pa.

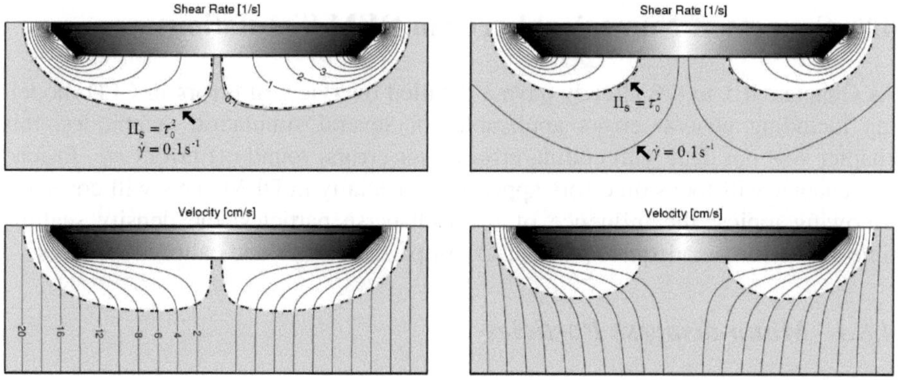

Fig. 4.7 Simulation results when using Eq. (22) with δ = 0.002 (to the left) and δ = 0.1 (to the right)

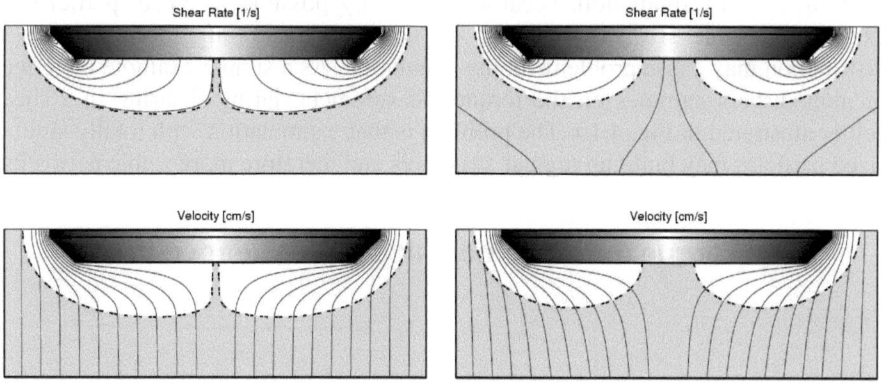

Fig. 4.8 Simulation results when using Eq. (23) with m = 80 (to the left) and m = 5 (to the right)

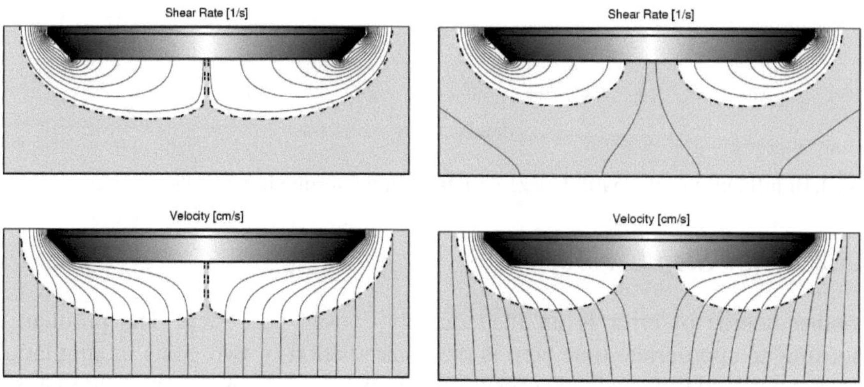

Fig. 4.9 Simulation results when using Eqs. (24) and (25) η_{max} = 104 Pa (left) and η_{max} = 500 Pa (right)

4.8 Sources of Numerical Error in DEM Simulations

As Chapters 4.1 to 4.6 already gave a detailed overview of errors in CFD modeling including general errors applicable for several simulation techniques, this chapter will not deal with coding errors, user errors, round off errors *etc.* Instead this chapter will focus on errors appearing especially in DEM. This will cover the following topics: the influence of mono-disperse particles, the density scaling, time-step errors, inadequate particle size or particle shape and calibration errors.

4.8.1 Mono-disperse Particles

When talking of particles of the same size in reality, this usually means the particles are about the same size, but have a uniform or normal distribution in specified ranges. This fact shall not be forgotten when modeling mono-disperse particles in a DEM simulation, because it is easily possible to create particles of exactly the same size. This should be done very carefully, because such a distribution may cause problems in the simulation. As a simple example the effect of monodispers particles on the torque measurement during shearing in a shear cell is illustrated in Fig. 4.10. The problem is that a simulation with totally similar sized particles may build up regular structures and therefore more stable networks.

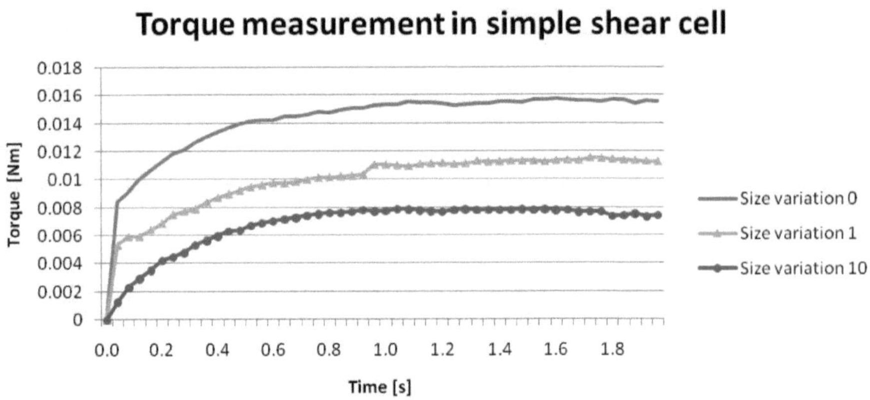

Fig. 4.10 Influence of the particle size variation on the torque in a simple shear cell

4.8.2 Time Step Errors

Another source of error is an inadequately chosen time step. The problem of choosing an appropriate time step is the contradiction of two goals in simulation: fast computation time and exact results. The goal is to choose a time step that is as large as possible to reduce the simulation time as much as possible and on the other hand is small enough not to significantly falsify the simulation results. In case of an oversized time step, enormous forces at the contacts may occur and therefore will make the simulation unstable. When particles move with a high speed and the time step is increased too much, particles can even cut through wall elements without touching them. If the forces during the impact are of interest, the time steps have to be chosen smaller. To capture the process of the impact forces, there have to be several sampling points during the impact. This problem of choosing an adequate time step is of course well known and there are different approaches to solve this problem implemented in DEM software. One approach is the automated time step calculation, using a critical time step. For the simple case this critical time step is defined by

$$T = \sqrt{\frac{m}{k}} \tag{26}$$

where m is the mass of a particle and k is its stiffness. Another approach for detecting an appropriate time step is based on the Rayleigh wave propagation:

$$T_R = \frac{\pi R \sqrt{\rho / G}}{0.1631\nu + 0.8766} \tag{27}$$

where R is the particle's radius, ρ is the density, G is the shear modulus and ν is the Poisson's ratio. These implementations are very useful and simplify the work for the user enormous. Nevertheless the suggested time steps are not always applicable and the right safe factors have to be chosen to make the simulation work stable on one hand and not slow down unnecessarily on the other. In addition to this basic problem, these time step calculations are based on default contact models, but when trying to simulate SCC these models only work hardly. Therefore user defined models are used, where the time step calculation is even more difficult to handle.

4.8.3 Density Scaling Errors

An issue related to the topic of the last paragraph is the problem of errors caused by density scaling. Density scaling is a technique offered in some DEM programs to set the time step to a fixed value. The goal is to reduce the simulation time, by increasing the time step, trying to avoid the problems described in the above section. Instead of just increasing the time step, the inertial masses of the particles are multiplied by a factor to fulfill the criteria for the critical time step, which is computed in dependency to the masses. The modified inertial masses have no effect on the gravitational mass. This technique is used when the interest of the user lies in the steady state of a simulation. The problem is, the simulation will only produce correct results, when the simulation is path independent. The problem is to estimate whether a simulation is path dependent or not.

4.8.4 Calibration Errors

An important part of simulations in DEM is the calibration of the micro parameters. These parameters usually cannot just be measured at the real material, but have to be calibrated by executing a simple reference experiment, covering the relevant behavior of the material. This experiment is first processed with the real material and afterwards modeled within the simulation. The micro parameters in the simulation are adjusted iteratively until the behavior fits the input. This process is difficult and empirical and therefore a significant source of error. Beyond this there are two more problems in calibration. First DEM is using abstract particle shape and often abstract particle sizes. These differences have to be represented by micro parameters as well, but it is not always granted that this simplification is appropriate. Second the calibration experiment usually only covers a special range of boundary conditions. Even when processing several of these experiments, it is very difficult to cover all conditions that can occur in the real process to be simulated. Therefore it is not granted that the simulated material behaves the same way as the real material would do, when special boundary conditions occur. The goal is to capture the essential properties within a certain scope of application using a multipoint calibration. As an example: may a model be given, representing the flow properties at low flow velocities adequately. In the case of high velocities this model may be defective, because it is out of the scope of application. Also the forces acting on the material under compression do not have to be correct, when the model is not calibrated for such conditions in advance.

4.8.5 Particle Size

Using the incorrect particle size is also a possible source of error within DEM simulations. In similarity to the time step issue, the particle size is influencing both the computation time and the modeling accuracy. Often the size of real particles is too small to be modeled in the simulation, because the amount of particles would be very large and this would increase the computation time dramatically. Therefore the particle size is increased to reach an appropriate computation speed. On the other hand the particle size is supposed to be small enough to model all essential phenomena within the simulation. A simple case in which problems occur arises when particle sizes exceed critical distances given by the boundary conditions. In the case of modeling SCC filling of a formwork with reinforcement, the maximum particle size should be of the same size than real aggregates, when trying to model blocking. Beside such restrictions that can be obtained with little effort, particles with larger diameters may show a different behavior in general, *e.g.* concerning inertia effects. As a very simple example the angle of repose may be brought up. If the particle size is too large in comparison to the volume to be modeled, the size effect will dominate the result of the simulation instead of the material properties (see Fig. 4.11 below).

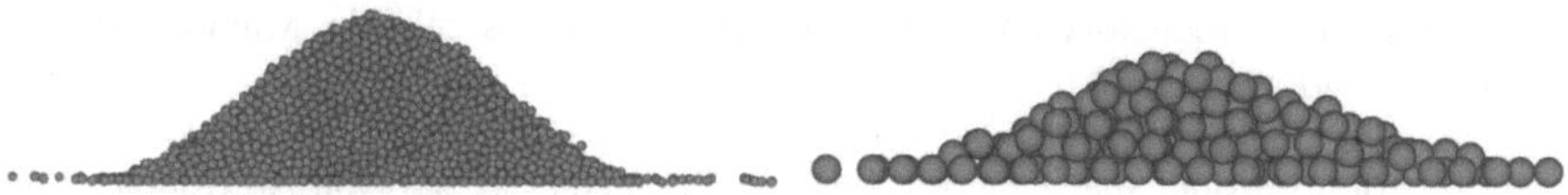

Fig. 4.11 Comparison of the angle of repose of a material with the same particle properties, but different particle sizes

4.8.6 Particle Shape

The basic shape of particles within the DEM is an ideal sphere, what makes the contact detection easy and enables a fast computation speed. Because of this simplification of the real shape, among other effects, the particles are rolling easily on plane surfaces, while a real gravel grain does not. To avoid this effect the rolling friction coefficient can be increased or the degrees of rolling can even be completely blocked. Both approaches have side effects on the material behavior. A more realistic mapping of the real particle geometry can be achieved by creating a complex particle (clump) by putting several spheres together, that have a fixed position relative to each other. The downside of this approach is the increasing computation time.

References

[1] AIAA, Guide for the Certification and Validation of Computational Fluid Dynamics Simulations. AIAA Guide G-077-1998 (1998)

[2] Casey, M., Wintergerste, T. (eds.): ERCOFTAC (2000), Best Practice Guidelines, Version 1.0. ERCOFTAC Special Interest Group on Quality and Trust Industrial CFD (2000)

[3] Versteeg, H.K., Malalasekera, W.: An introduction to computational fluid dynamics – the finite volume method, 2nd edn. Pearson Education Limited, England (2007)

[4] Oberkampf, W.L., Trucano, T.G.: Verification and Validation in Computational Fluid Dynamics. Prog. Aerosp. Sci. 38, 209–272 (2002)

[5] Mase, G.E.: Schaums Outline Series: Theory and Problems of Continuum Mechanics. McGraw–Hill Inc., USA (1970)

[6] Malvern, L.E.: Introduction to the Mechanics of Continuous Medium. Prentice-Hall Inc., New Jersey (1969)

[7] Edwards Jr., C.H., Penney, D.E.: Calculus and Analytical Geometry, 2nd edn. Prentice-Hall, Inc., USA (1986)

[8] Langtangen, H.P.: Computational Partial Differential Equations, Numerical Methods and Diffpack Programming. Lecture Notes in Computational Science and Engineering. Springer, Berlin (1999)

[9] Wallevik, J.E.: Rheology of Particle Suspensions - Fresh Concrete, Mortar and Cement Paste with Various Types of Lignosulfonates (Ph.D.-thesis); Department of Structural Engineering, The Norwegian University of Science and Technology (2003), `http://www.diva-portal.org/` ISBN 82-471-5566-4, ISSN 0809-103X

[10] Kolman, B.: Introductory linear algebra with applications, 5th edn. Maxwell Macmillan International (1993)
[11] Wallevik, J.E.: Minimizing end–effects in the coaxial cylinders viscometer: Viscoplastic flow inside the ConTec BML Viscometer 3. J. Non-Newtonian Fluid Mech. 155, 116–123 (2008)
[12] Anderson, J.D.: Computational Fluid Dynamics, The Basics with Applications. McGraw-Hill, Inc., USA (1995)
[13] Fletcher, C.A.J.: Computational Techniques for Fluid Dynamics, 2nd edn. Springer Series in Computational Physics, vol. I. Springer, Germany (1990)
[14] Tattersall, G.H., Bloomer, S.J.: Further development of the two–point test for workability and extension of its range. Mag. Concr. Res. 31(109), 202–210 (1979)
[15] Tattersall, G.H., Banfill, P.F.G.: The Rheology of Fresh Concrete. Pitman Books Limited, Great Britain (1983)
[16] Tattersall, G.H.: Workability and Quality Control of Concrete. E & FN Spon, Great Britain (1991)
[17] Domone, P.L.J., Yongmo, X., Banfill, P.F.G.: Developments of the two–point workability test for high–performance concrete. Mag. Concr. Res. 51(3), 171–179 (1999)
[18] Wallevik, J.E.: Rheological properties of cement paste: thixotropic behavior and structural breakdown. Cement Concr. Res. 39, 14–29 (2009)
[19] Oldroyd, J.G.: A Rational Formulation of the Equations of Plastic Flow for a Bingham Solid. Proc. Camb. Philos. Soc. 43, 100–105 (1947)
[20] Oldroyd, J.G.: Two-Dimensional Plastic Flow of a Bingham Solid. Proc. Camb. Philos. Soc. 43, 383–395 (1947)
[21] Barnes, H.A., Hutton, J.F., Walters, K.: An Introduction to Rheology. Elsevier Science, Amsterdam (1989)
[22] Irgens, F.: Continuum Mechanics. Springer, Berlin (2008)
[23] Tanner, R.I., Walters, K.: Rheology: An Historical Perspective. Elsevier Science B.V., Netherlands (1998)
[24] Bercovier, M., Engelman, M.: A finite element method for incompressible non–Newtonian flows. J. Comput. Phys. 36, 313–326 (1980)
[25] Taylor, A.J., Wilson, S.D.R.: Conduit flow of an incompressible, yield–stress fluid. J. Rheol. 41(1), 93–101 (1997)
[26] Burgos, G.R., Alexandrou, A.N., Entov, V.: On the determination of yield surfaces in Herschel–Bulkley fluids. J. Rheol. 43(3), 463–483 (1999)
[27] Papanastasiou, T.C.: Flow of material with yield. J. Rheol. 31, 385–404 (1987)
[28] O'Donovan, E.J., Tanner, R.I.: Numerical study of the Bingham squeeze film problem. J. Non-Newtoninan Fluid Mech. 15, 75–83 (1984)
[29] Wallevik, J.E.: Development of Parallel Plate-Based Measuring System for the ConTec Viscometer. In: Proceedings of the 3rd International RILEM Symposium on Rheology of Cement Suspensions such as Fresh Concrete, August 19-21, RILEM Publications S.A.R.L., Reykjavik (2009) ISBN: 978-2-35158-091-2

Chapter 5
Advanced Methods and Future Perspectives

Ksenija Vasilic[1], Mette Geiker[2], Jesper Hattel[2], Laetitia Martinie[3], Nicos Martys[4], Nicolas Roussel[3], and Jon Spangenberg[2]

[1] BAM, Bundesanstalt für Materialforschung und–prüfung, Germany
[2] DTU, Danmarks Tekniske Universitet, Denmark
[3] Université Paris Est, IFSTTAR, Paris, France
[4] NIST, National Institute of Standards and Technology, Maryland, USA

5.1 Introduction

The one-phase methods described in Chapter 2 were shown to be able to predict casting to some extent, but could not depict segregation, sedimentation and blockage occurring during flow. On the other hand, the distinct element methods described in Chapter 3 did not take into account the presence of two phases in the system and describes concrete as distinct elements interacting through more or less complex laws. A reliable numerical model of a multiphase material behaviour shall take into account both phases (solid and liquid). From the numerical point of view, concrete flow shall be seen therefore as the free surface flow of a highly-concentrated suspension of rigid grains.

In many other engineering fields, suspension flows have been successfully simulated and sophisticated methods and models have been developed. These models though usually do not involve such a wide range of particles and the volume fraction of the solid phase is far lower than in concrete. Therefore, although useful, the experience from the other fields cannot be directly applied to the simulations of concrete flow.

Common approach in the field of concrete is to consider concrete as a two phase suspension (*cf.* Fig. 5.1): the "liquid phase" made of either cement paste or mortar and the dispersed phase made of the coarsest particles.

As the techniques in this field are very recent and strongly differs one from another, we choose in this chapter to give some examples of numerical simulations developed by various authors in order to simulate the presence of grains or fibres in a non-Newtonian cement matrix. It shall be kept in mind that these are advanced techniques still under development. They require, for most of them, high computation capacity but represent, in the opinion of the authors, the future of numerical simulations of fresh concrete flow.

N. Roussel and A. Gram (eds.), *Simulation of Fresh Concrete Flow*, RILEM State-of-the-Art Reports 15,
DOI: 10.1007/978-94-017-8884-7_5, © RILEM 2014

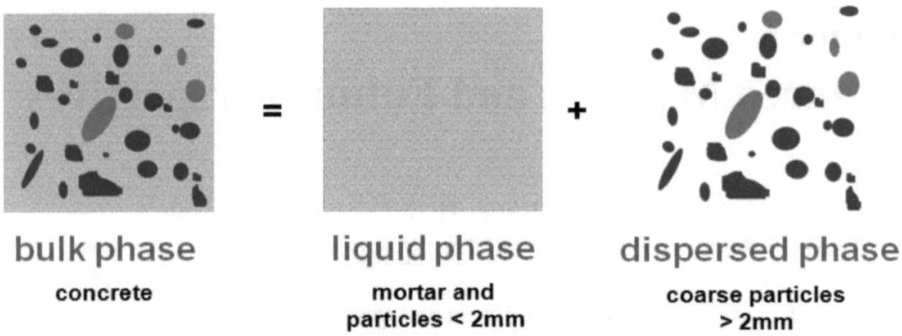

Fig. 5.1 Concrete represented as a two phase suspension

5.2 Case Studies

5.2.1 FEMLIP Method from EC Nantes

Based on the Material Point Method originally developed by Sulsky and Schreyer [1] and belonging to the broad family of Particle-In-Cell method, Moresi *et al.* [2] have recently developed a new hybrid FEM (FEMLIP), where material points and computational points are decoupled. This computational method makes use of a combination of Lagrangian and Eulerian.

The original idea was to keep using FEM well known around the scientific community for its robustness and versatility and modify it to account for infinitely large deformation. The main drawback of Lagrangian FEM for large deformation applications is the element distortion which, once it has reached a critical state, yields inefficiency and inaccuracy of the element-wise integration scheme. Therefore one needs to disconnect material and computational points in order to keep nicely shaped finite elements even for extremely large material deformation. The retained approach uses an Eulerian finite element grid (fixed) as a computational set of points and a set of Lagrangian particles embedded in the mesh which are used as integration points for any given configuration (Fig. 5.2(b)). Material properties are initially set on particles (Fig. 5.2(a)).

Once nodal unknowns (velocity and pressure for incompressible materials) are computed for any configuration using element-wise integration scheme over particles, the velocity field is interpolated to the particles. In contrary to Gaussian points in classical FEM, shape functions need to be recalculated for each particle configuration as their position are changing through time within elements. Particles positions are updated according to their velocity (Fig. 5.2(c)). All time dependent properties stored on particles move with the particles and must be rotated if they are tensor variables (*e.g.* elastic stress tensor). At the end of the computational step the grid contains no information at all (it is a pure computational grid) and one can choose a totally different mesh for the new particle configuration. The grid is usually kept fixed except in the case of moving boundary conditions.

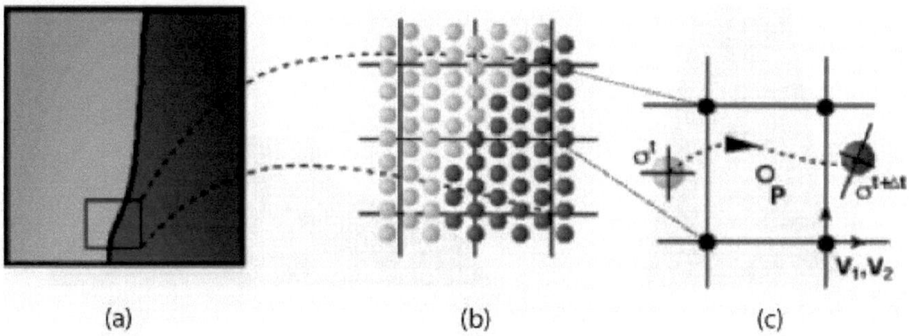

Fig. 5.2 Schematic representation of FEMLIP (a) Interfaces are set by applying different behaviour to material in space (b) A set of Lagrangian particles are used as integration points and to track material properties (c) Particles are moved according to nodal velocities and history dependant variables are stored on particles

The capacity of FEMLIP to simulate very large deformation processes with interfaces and free surfaces has been demonstrated for a wide range of material properties in several types of applications including concrete flow in forms. When modelling concrete as a heterogeneous material made of mortar and aggregate (see Fig. 5.3 (top)), the scale should be smaller than the form scale since aggregates must be discretized by several finite elements in order to be properly modelled as rigid compared to mortar. Dufour and Pijaudier-Cabot [3] applied this method on two different types of concrete (self-compacting and ordinary concrete). They fitted the two Bingham's parameters on experimental results from a simple slump test with flow time measurement (see Fig. 5.3).

5.2.2 Two-Phase Model from IBAC and IVT

In [4] Modigell *et al.* suggested a two phase model for simulation of SCC flow. Here, a so-called pseudo continuum approach is used [5] in which the suspension is regarded as a mixture of two phases: the continuous liquid matrix and the dispersed solid phase, modelled as a pseudo-liquid. The constitutive equations are solved for the solid phase and the "quasi" liquid phase, consisting of the liquid phase with suspended non-interacting particles. In the momentum conservation equation the interaction between solid and liquid phase is modelled by Darcy's law [6].

The viscosity of the bulk phase, which is a mixture of the solid and liquid phase, is described as follows:

$$\eta_B(f_S, \dot{\gamma}, t) = \left(\frac{\tau_0}{\dot{\gamma}} + k\dot{\gamma}^{m-1} \right) \cdot \kappa + \eta_L \tag{1}$$

where indices L and B denote the liquid and the bulk phase, respectively.

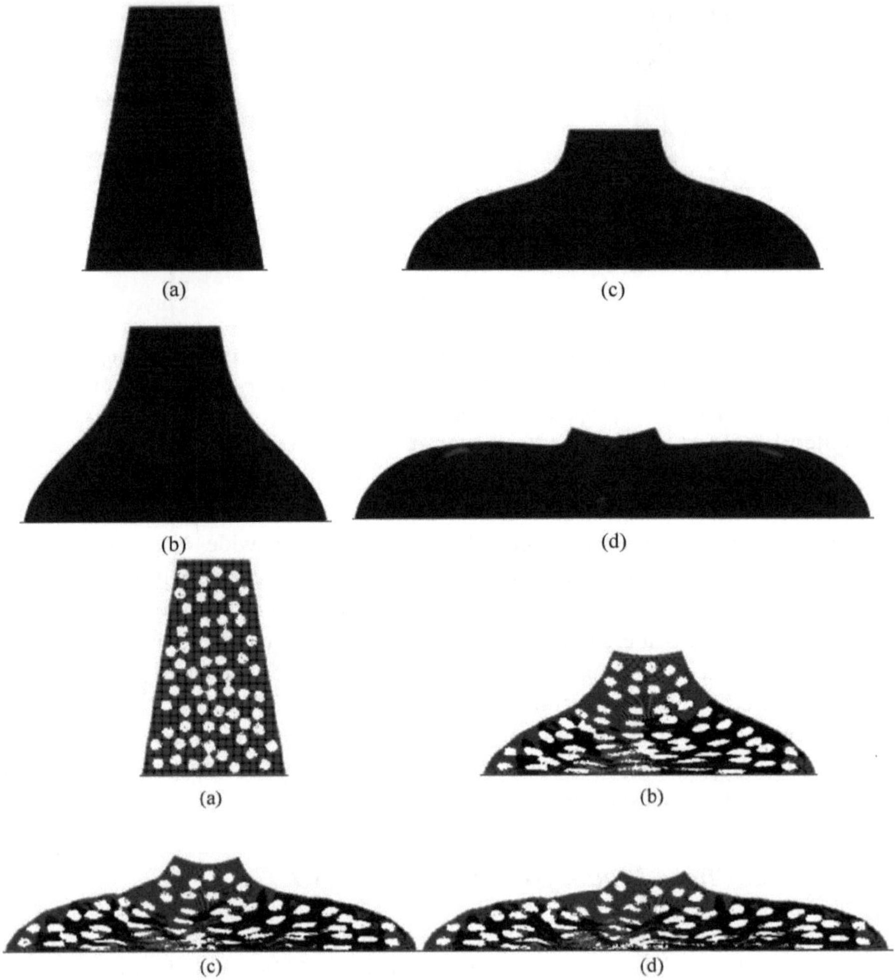

Fig. 5.3 (top) Numerical simulation of the slump flow test using a homogeneous approach (bottom) Numerical simulation of the slump flow test using a heterogeneous approach, from [3]

The degree of agglomeration κ describes the kinetics of the structural built-up of a material and is related to the current suspension structure. In a fully structured state, κ is assumed to be unity and in a fully deagglomerated state it is assumed to be zero. The kinetics of the structural change is modelled as a first order equilibrium reaction:

$$\frac{D\kappa}{Dt} = C(\kappa_e - \kappa) \tag{2}$$

with C being the reaction rate coefficient and κ_e the equilibrium value of the structural parameter for the applied stress. All the parameters strongly depend on the solid content. This solid content changes locally as described in the solid fraction continuity equation:

$$\frac{Df_s}{Dt} + \left(\nabla \cdot \underline{u}_S\right) f_s = 0 \tag{3}$$

The suggested two-phase model, which combines the pseudo-solid and thixotropy approach, has previously bee used by authors to model the complex suspension flows [6]. The flow equations shown above, coupled with the constitutive equations for the bulk and liquid phase, as well as the free surface equation, are implemented into the CFD software PETERA©. The numerical code PETERA© is based on the Lagrange-Galerkin approach that uses a special discretization of the Lagrangian material derivative along particle trajectories with a Galerkin Finite Element Method [7]. The numerical model uses an analytical solution of rate equations along segments of particle trajectories as well as backward particles trajectories tracking in time.

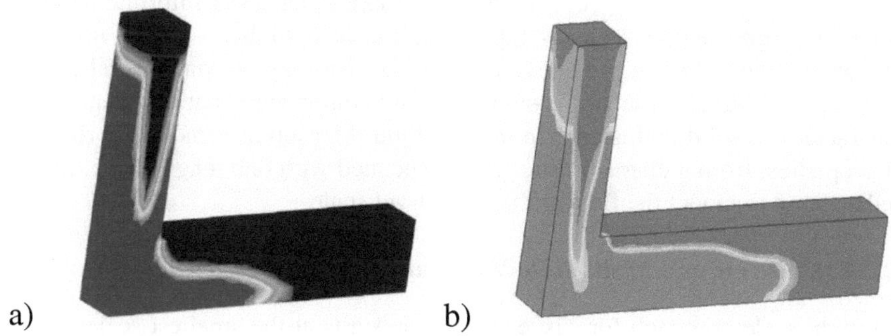

Fig. 5.4 Numerical results of the volume fraction of the fluid and air after 0.3 s for the one-phase simulation (a) and the two phase simulation (b)

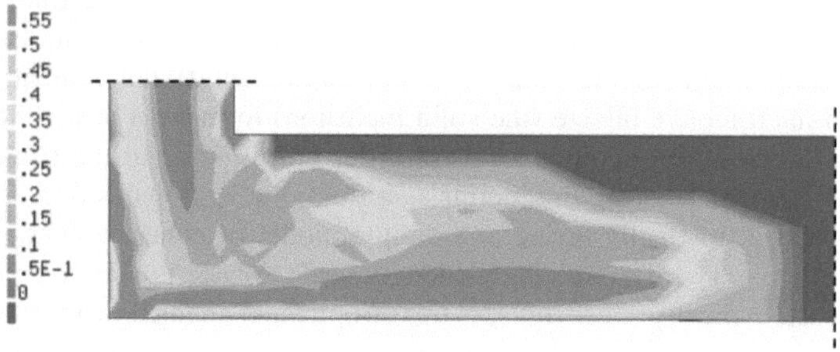

Fig. 5.5 Numerical result of the two-phase simulation: detail of the solid fraction distribution in the fluid after 0.3 s. Occurrence of higher solid content and phase segregation is noticeable.

In the article, the authors compared one-phase and two-phase simulations of L-Shape experiment. The PETERA© code is used to perform the two-phase simulation, while the one-phase simulation is conducted applying the FLUENT 6.2.16 software. They showed that there is a difference in the flow front between these two simulations, and it is believed that the difference is a result of the particle segregation which occurs in the material, and which influences the solid fraction distribution as well as the material viscosity. The phase segregation occurs in the horizontal duct where dispersed coarse particles show a tendency to settle (high solid content layer in lower part of the horizontal duct). The segregation also takes place on the corner wall of the box, when the flow direction changes from vertical to horizontal channel.

5.2.3 Dissipative Particle Dynamics from NIST

In [8], Martys applied multi scale modelling of suspensions to cement based materials. A complete description of concrete from the cement particles to coarse aggregate is impossible with any computer model [9] since accounting for broad particle size and shape distributions exceeds the computational limits of even the best super computers. Hence, a multi-scale model that takes information corresponding to one length scale as input into the next higher scale is needed to develop a comprehensive predictive tool for the rheology of concrete. The multi-scale approach divides the suspension into fluid components that are identified by a characteristic solid inclusion size in each fluid. Measured or modelled rheological properties, from a characteristic fluid associated with one length scale, may be used as input to model the flow at the next larger scale.

5.2.3.1 Concrete as a Multi-scale Material

In a multi-scale approach for suspensions, one starts at the smallest scales, defining a suspension by a matrix or embedding fluid that contains solid inclusions. This suspension now becomes the matrix fluid for larger solid inclusions (typically ten times the size of the particles in the matrix fluid). This process can again be repeated over and over until we reach the final macroscopic fluid of interest. The author considers a simple multi-scale model of concrete. Starting with the smallest scale one combines water (the embedding fluid) and cement particles of order ten micrometers in size (the solid inclusion) to make a new fluid called cement paste. Now, if one adds sand (of order mm in size) to the cement paste, a mortar is formed. Treating the mortar as a fluid, the aggregates (cm size in scale) are then added, to finally arrive at a simple model of concrete. The matrix fluid properties are very different at each length scale. Water is Newtonian, while the cement paste typically shear thins, is thixotropic, and changes with time becoming visco-elastic as it forms a gel and undergoes hydration. Mortar is non-Newtonian and shear thins. Concrete is non-Newtonian and is known to jam as it flows through narrowly placed rebars [8]. Of course, whatever changes that take place on one scale will alter the behaviour at large scales.

Clearly this model is an oversimplification of concrete. First, the particle sizes at each selected length scale can easily vary by an order of magnitude or more. Except for, perhaps, a gap graded concrete, the size variation of inclusions can overlap so that it may not always be clear how to separate scales. There is the further complication in that a realistic description of a cement paste, mortar, or concrete requires a correct model for non-Newtonian fluid behaviour, *e.g.* fluids with a yield stress or shear thinning property, usually described as a Bingham fluid or a Herschel Bulkley fluid. Therefore, making a transition from cement paste to mortar and then to concrete means changing the nature of the fluid, not simply the size of the suspended inclusion.

5.2.3.2 Computational Models

To study fluid flow behaviour corresponding to the different length scales, Martys has developed several computational models based on Dissipative Particle Hydrodynamics (DPD) [8], Smoothed Particle Hydrodynamics (SPH) [10] and alternate approaches like Lattice Boltzmann [11]. Both DPD and SPH can be thought of as a Lagrangian formulation of the Navier Stokes equations. Although not correct historically, DPD can be thought of as a simplified version of SPH that is useful when the particles are of order a micrometer in size where Brownian motion is important. These approaches are very similar in structure in that a fluid phase is represented by a set of points (Fig. 5.6) that carry the fluid properties. Neighbouring particles are allowed to interact according to certain rules that conserve mass and momentums in such a way that the Navier Stokes equations are recovered. Solid inclusions are represented by "freezing" a set of the fluid particles so that they move together as a rigid body [12].

An important capability of SPH is that it is able to describe the flow of non-Newtonian fluids. Having such a capability is crucial as shear rates can vary quite broadly in a suspension because the space between inclusions can differ significantly. As a consequence, depending on the viscosity vs. shear rate relation for the matrix fluid, extremely heterogeneous values of local viscosity throughout the suspension may exist. The author developed a model, based on the SPH approach, of suspensions where the matrix or embedding fluid may be non-Newtonian. This model is a Lagrangian formulation of the Generalized Navier-Stokes equations (GNS) [13]. The Lagrangian representation of the continuity equation and the GNS is given by

$$\frac{\partial \rho}{\partial t} = -\rho \nabla \cdot v \tag{4}$$

$$\rho \frac{\partial v_i}{\partial t} = -\frac{\partial P}{\partial x_i} + \frac{\partial}{\partial x_k}\left[\mu\left(\frac{\partial v_i}{\partial x_k} + \frac{\partial v_k}{\partial x_i} - \frac{2}{3}\delta_{ik}\nabla \cdot v\right)\right] + \frac{\partial}{\partial x_i}(\zeta \nabla \cdot v) \tag{5}$$

respectively. Here ρ is the fluid density, μ is the shear viscosity, ζ is the bulk viscosity, v is the fluid velocity and, P is the pressure. Because the matrix fluid is

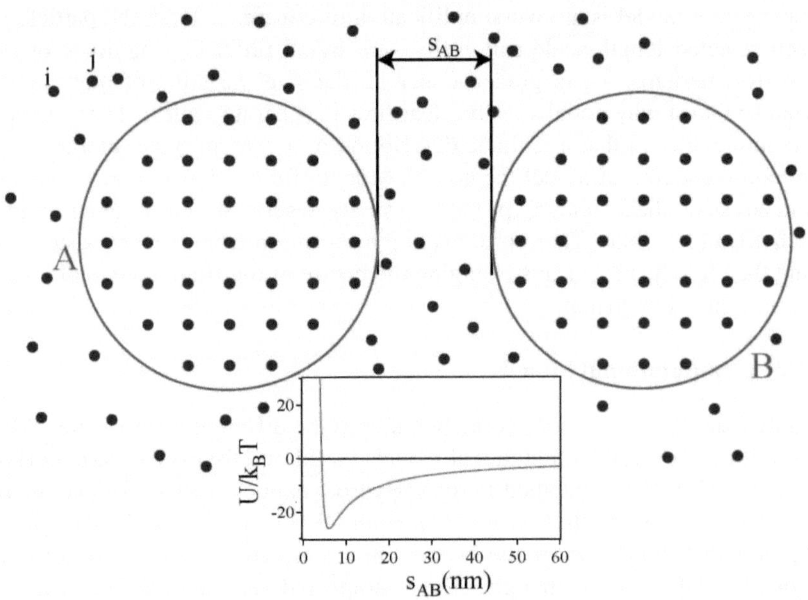

Fig. 5.6 Illustration of simulated fluid suspension. The fluid suspension is represented by DPD or SPH particles (in blue). The enclosed regions A and B represent solid inclusions. The assembly of particles, enclosed in these regions, is subject to constraints such that they move together as a rigid body. An interaction potential, U, may be introduced that depends on the nearest surface to surface distance (SAB) between inclusions A and B.

non-Newtonian, the viscosity can vary from point to point hence the gradient of the viscosity must be evaluated in the momentum equation. The SPH formalism allows for the spatial discretization of the fluid into moving particles that carry information about the momentum, density and, if needed, temperature of the fluid. From this information derivatives of fluid variables can be constructed to represent the GNS.

This approach has been validated for a variety of flow scenarios. A non-trivial example is given in Figs. 5.7, 5.8, 5.9, and 5.10, which compare experimental results to simulation. In this case, a constitutive relation between viscosity and shear rate, which was determined from experimental data, was entered into the simulation code to represent a shear thinning matrix fluid. Adding different volume fractions of spheres to this non-Newtonian embedding fluid, the flow behaviour of this suspension was compared to experimental measurements. Excellent agreement was obtained between simulation prediction and experiment measurements of viscosity (Fig. 5.9). This approach was also tested, with reasonable success, for non spherical shape inclusions that are typically found in cement based materials (Fig. 5.8).

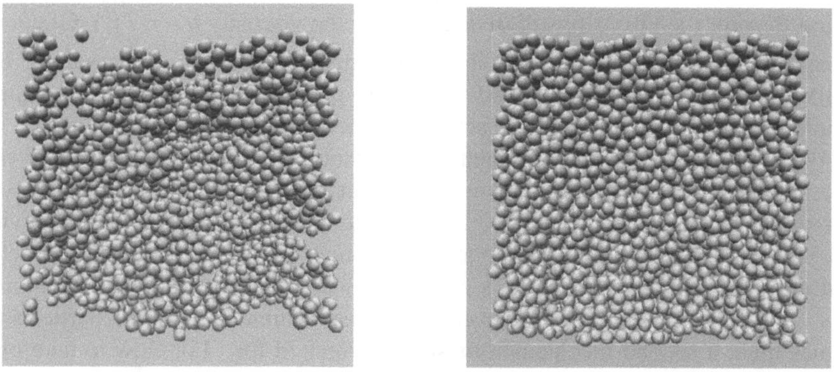

Fig. 5.7 Hard sphere suspensions subject to shear in simulation. The solid volume fractions are 20 % and 50 %.

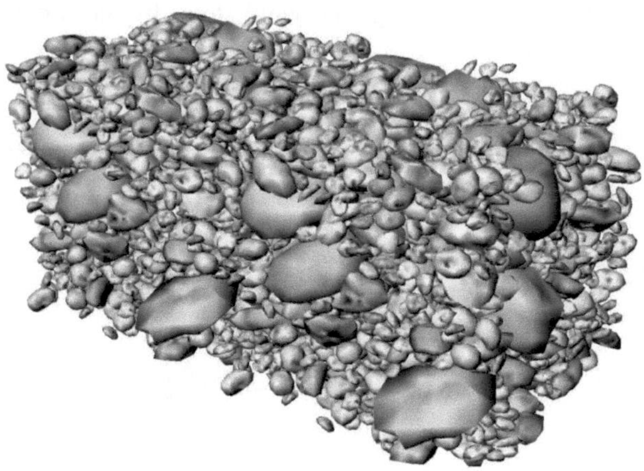

Fig. 5.8 Mortar suspension composed of sand particles. The sand particles are based on X-Ray micro tomography images of sand that can be incorporated into the simulation code.

5.2.3.3 Some Fundamental Insights into Yield Stress

To investigate the shear rate dependence of yield stress a set of simulations were carried out using strongly attractive spheres systems subject to Brownian motion. For such systems we found that, in the limit of zero shear rate, the actual yield stress, as measured by the maximum stress, σ_m, during stress growth actually goes to zero with decreasing shear rate, where $\sigma_m \sim \dot{\gamma}^n$ and n = .25± .05. This is a consequence of thermal fluctuations or Brownian motion playing a bigger roll as the shear rate is decreased. An interesting result was that despite the yield stress going to zero, the viscosity still diverged (note the viscosity $\mu \sim \sigma / \dot{\gamma}$). Of course, measurements at such very low shear rates also imply long times scales. It is unlikely that this behaviour would be seen in cement paste as chemical changes would actually be the controlling factor over such long time scales.

We also considered fundamental mechanisms that lead to a measured yield stress behaviour. Scaling arguments show that yield stress is proportional to the maximum force between particles. Here one assumes that the yielding behaviour is a consequence of pulling apart particles to break up a previous formed structure and allow flow. This conjecture is consistent with our simulations in the limit of low solid volume fraction. However, when the volume fraction of particles becomes high, a second mechanism must be accounted for. For flow to take place the inclusions must push against each other, or displace their neighbours. Under shear, this process usually takes place in compression. So we find there are two contributions to yield stress; the breaking up of bonds and the displacing of neighbouring inclusions. As a result, simply modifying the maximum inter particle force by some factor does not change the yield stress by the same factor in high volume fraction suspensions.

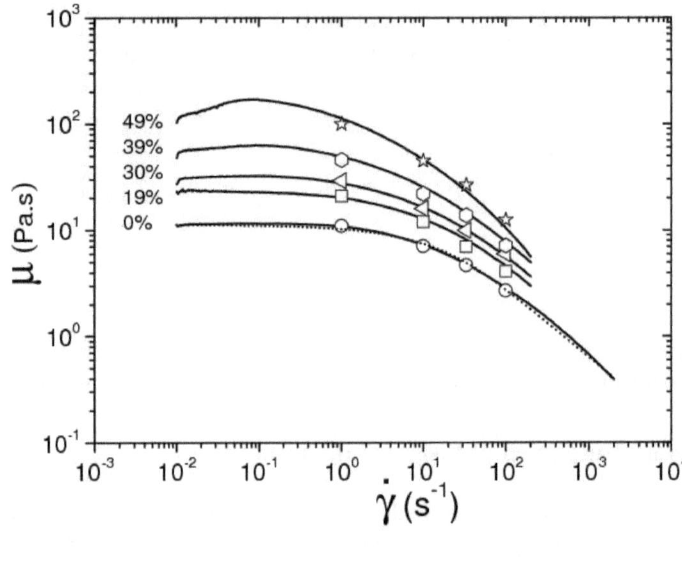

Fig. 5.9 Comparison of experimental measurements (solids lines) of viscosity as a function of shear rate for a shear thinning fluid (0 %) and suspensions composed of silica spherical particles at designated volume fraction volume fractions (19 %, 30 %, 39 % and 49 %) with simulation data (open symbols). Based on sample preparation, experimental results could vary about 10 % from that shown in the figure.

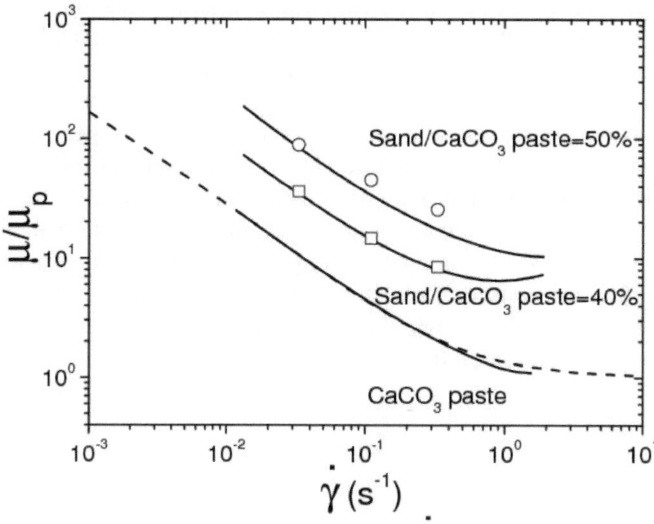

Fig. 5.10 Comparison of simulation (open symbols) results to experimental measurements (solid lines) of relative viscosity vs. shear rate for a suspension composed of sand embedded in a calcium carbonate paste. The matrix fluid (bottom solid line) and an approximate fit (dashed line) to the data are also shown. Data for volume fractions 40 % and 50 % are shown. Based on sample preparation, experimental results could vary about 10 % from that shown in the figure.

5.2.3.4 Insights to the Effect of Particle Sizes and Shapes

Furthermore, numerical studies on the influence of the particle size and shape variation are conducted. It has been previously shown that replacing crushed aggregate with spherical shapes can decrease the viscosity by a factor of approximately two [8]. An interesting application of this idea is the case of replacing cement with smaller fly ash particles. It is well known that fly ash can improve the workability of cement based materials. This improvement is largely attributed to the spherical nature of the fly ash, which is thought to produce a ball bearing-like effect to help flow.

The simulations (*cf.* Fig. 5.9) have shown that while there may be some ball bearing effect, the main improvement is due to the breaking up of larger scale structures that develop as the cement particles are sheared. The angular or faceted shape of the cement particles can lend itself to the formation of chains of particles that interact in compression, which has a tendency to increase the viscosity. Adding the smaller fly ash breaks up the chains so that less force is transmitted in this fashion and as a result the viscosity was lower. As a variation of this study, the simulation was repeated with the same volume fraction of cement particles replaced with spherical ones having the same individual volume of those replaced.

In this case, although there was some improvement, it was not as dramatic as that seen in experiments [14]. An important point is that in this simulation it was assumed that the particles did not have any attractive interaction. Only a very short distance repulsive force was used to make sure the particles did not overlap over the course of the simulation. So this simulation could roughly be thought of as describing a system where there is a sufficient dosage of super plasticizer that prevents agglomeration of particles through Van der Waals or other attractive interactions. If one assumes a strong attractive interaction between the cement particles, the fly ash does not agglomerate nor is attractive to the cement, then the cement is prevented from gelling or forming part of a network. This can significantly reduce the viscosity or yield stress of the fluid as little stress is supported over large scales. Understanding these results, from a fundamental view, still remains an important research topic.

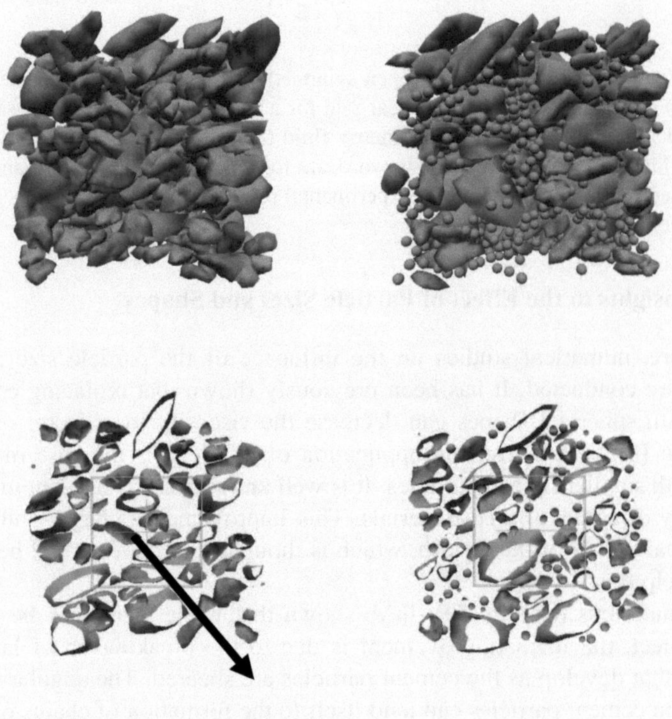

Fig. 5.11 Cement paste suspension without and with ultra fine fly ash. "Slices" of image taken after undergoing shear that reveals that the ultra fine fly ash may disrupt networks that allow for transmission of forces along chains (indicated by arrow) of particles. As a result the viscosity and yield stress is lower for these systems.

5.2.4 *Prediction of Dynamic Segregation from DTU*

During casting of SCC and before setting heterogeneities can be induced if the cement paste is too fluid to carry the aggregates. A heterogenic aggregate distribution may lead to significant variations of the concrete properties and hence to a potential decrease of the structural performance. Recently, [15] have shown that it is possible to use a computational fluid dynamics calculation of the casting process coupled with numerical modelling of the segregation of the coarsest particles to produce a map of properties in the hardened state of a structural element.

An example of such a prediction of dynamic segregation (flow induced heterogeneities) by numerical simulations was presented in [15] using the software FLOW3D. Fig. 5.12 shows the final distribution of the particles when simulating the SCC casting of a 3 meters long and 30 cm high beam. The forces taken into account to affect the particles are the buoyancy, gravity and drag. The position of the particles is calculated explicitly with a one way momentum coupling between the continuous phase and the particles (*i.e.* the particles "feel" the continuous phase but not vice versa). Also, no interaction between the particles is assumed.

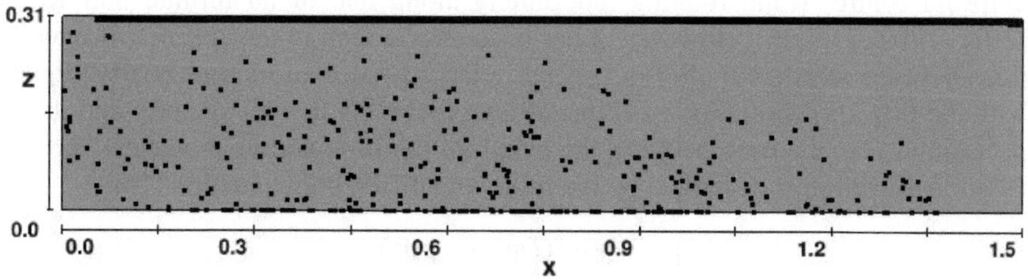

Fig. 5.12 Numerical simulation of the heterogenic aggregate distribution phenomenon after casting of a3 m long beam. SCC modelled as particles suspended in a continuous phase. Continuous phase modelled with a yield stress of 50 and a plastic viscosity of 50 Pa s. Aggregates modelled as spherical particles with a diameter of 15 mm. [15]

A more accurate prediction of dynamic segregation in SCC could be obtained by modelling a full momentum coupling between particles and continuous phase. However, the calculation time of such a coupling is very extensive even with today's (2010) multiple CPU calculation possibilities.

5.2.5 *Fibre Orientation Modelling*

5.2.5.1 Industrial Flow

The orientation of a population of fibres reinforcing a cementitious material strongly influences the material rheological behaviour, and, after setting, affects its mechanical properties in its hardened state. It thus appears essential to predict this orientation in order to optimize mix design and mechanical properties. In industrial conditions, numerical simulations are required. As, many constraints due to casting can influence the macroscopic orientation. Indeed, first, civil engineering structures involve very large volumes of yield stress fluids reinforced with very high quantities of fibres. For example, 100 litres of materials reinforced with 1 % (in volume) of fibres typically used in civil engineering lead to more than 10^6 fibres (10 mm length and 0.2 mm diameter). Then, these materials flow through complex geometries, with most of time large interface areas with the formwork walls and free surfaces. Finally, the pouring process controls the pouring debit and sequence.

5.2.5.2 Background

The orientation process, first described by Jeffery [16] in 1922, is still of interest for many researchers, mainly in the case of Newtonian fluids [17-30]. Jeffery [16] solved the moment equation of a rotating ellipsoid immersed in an incompressible Newtonian fluid. He showed that a single rigid, force and torque free ellipsoid translates with a velocity equal to the one of the equivalent undisturbed fluid at the ellipsoid centre, while rotating indefinitely along one of an infinite number of Jeffery orbits [16, 26]. However, it has been shown that, in the absence of interactions between fibres, the ellipsoid spends a large portion of its time nearly aligned with the flow direction. For an ellipsoid represented by its unit vector p and its aspect ratio r (*i.e.* the ratio between the length of the fibre and its diameter), the orientation evolution is described by the following differential equation [16]:

$$\dot{\mathbf{p}} = W\mathbf{p} + \lambda(D\mathbf{p} - D : \mathbf{ppp}) \tag{6}$$

with W and D the vorticity tensor and the rate of strain tensor respectively for the flow unperturbed by the fibre. λ is a parameter representative of the fibre slenderness, $\lambda = (r^2 - 1)/(r^2 + 1)$. To predict orientation of cylindrical particles like fibres, the ellipsoid aspect ratio can be replaced in (6) by an effective aspect ratio r_e [26]. Experimental values of r_e / r are typically about 0.4-0.7 for high aspect ratios [19-22].

5.2.5.3 Aligned Fibre Assumption

The asymptotic case of an ellipsoid infinitely elongated is commonly used in literature to represent a cylindrical fibre and is called the aligned fibre approximation [28]. In this case, the fibre no longer rotates periodically in shear flow but approaches an equilibrium position oriented in the direction of the flow. Numerically, this assumption implies an infinite aspect ratio and thus a parameter λ approaching 1 in Eq. (6).

5.2.5.4 Interactions between Fibres

In order to represent interactions such as hydrodynamic interactions [31] between fibres in semi-dilute or concentrated materials, a complementary term is added to the Jeffery Eq. (6). Different approaches are used. The most common approach is the hydro-dynamically induced rotary diffusivity of isotropy [17, 32]. Indeed, the presence of fibres induces small disturbances in the velocity field. These disturbances cause the surrounding fibres to undergo some small rotations. Since these interactions are random, the small changes in the fibre orientation can be considered diffusive [18].

5.2.5.5 Yield Stress Effect

To locally apply the Jeffery Eq. (6) to a yield stress fluid, it is assumed that, in a volume proportional to the gyratory volume around the inclusion and centred on its inertia centre, the local apparent viscosity $\eta + \tau_0 / \dot{\gamma}$ is constant. It can be noticed that, for this assumption to be valid, the fibre length has to be a lot less than the flow characteristic size.

Literature provides only few experimental results in the case of yield stress fluids [18-28, 33-37]. The existence of a critical stress to overcome to induce flow creates areas which are not sheared, called plug-flow areas. Two orientation regimes can thus be observed. A) Outside the plug flow areas, deformations induced by the flow contribute to the fibre alignment as described by Jeffery, depending on the reinforced material rheological parameters and the fibre initial orientation. High deformation rates are generally localized at the interface with the mould due to non-slip boundary conditions. Thus, in addition to geometrical considerations and possible wall effect [38], the closer to the wall the fibre starts, the quicker it aligns with the flow. Then deformation rates decrease towards free surfaces or symmetrical plans, where stress due to the flow equals zero. B) Inside the plug-flow areas, the material behaves like a solid. Therefore, a fibre initially inside a plug-flow area keeps its initial orientation.

A simple dimensional approach shows that inside the sheared areas, fibre alignment with the flow is quasi instantaneous [39]. Thus, the main difficulty to be solved to predict fibre orientation in complex flows lies in the prediction of sheared/unsheared areas.

5.2.5.6 Multi-fibres Approach

Within the frame of this approach, the evolution of a finite group of slender fibres with isotropic initial orientations is considered to represent the macroscopic orientation process. At each time step, the evolution of the orientation of each fibre is directly deduced from the Jeffery equation with the diffusive term of interactions. A simple Euler explicit scheme of the first order is implemented in a CFD code Flow 3D ©. The macroscopic orientation is then calculated as the mean orientations of all fibres. The number of fibres chosen is motivated by both a good accuracy of the prediction and the computation time. Several configurations have been tested in [40], and it was found that 7 fibres distributed in a range of $\left[-\pi/2, \pi/2\right]$ initial angles provide the best compromise between the two requirements.

5.2.5.7 Application to Shear Flow between Two Parallel Walls

A fibre flowing through two infinite walls is submitted to a symmetrical shear stress field relatively to the median plan. The shear stress decreases from the wall interface to the symmetrical plane until reaching the material yield stress value at a critical distance from the wall. The central area is then translated with the fluid at the velocity of the interface with the sheared part.

From a mechanical point of view, there is an obvious interest in knowing the number of fibres crossing a given cracked section of a structural element. If we assume that the concentration of fibres can be considered as homogeneous (*i.e.* no static nor dynamic segregation of the fibres), this number is proportional to the total number of fibres per unit of volume through a parameter α called the orientation factor [38, 41-42]. This factor, expressed relatively to a direction $\vec{d}$, is derived from the analytical averaged projection of N fibres along this direction.

$$\alpha_{\vec{d}} = \frac{\sum_{k=1}^{N} \cos^2\theta_{\vec{d}}^k}{N} = \frac{\sum_{k=1}^{N} (p_{\vec{d}}^k)^2}{N} \tag{7}$$

with $\vec{d}$ the chosen direction, and $p_{\vec{d}}$ the fibre projection along direction $\vec{d}$. Values of this dimensionless factor vary in a range of [0, 1], with 1 for fibres perfectly aligned with $\vec{d}$, zero for fibres orthogonal to this direction, and equal to 0.5 for isotropic fibres.

The average orientation of steel fibres in a yield stress fluid flowing between two parallel plates can be computed. Results for steady state are presented on Fig. 5.13.

5.2.5.8 Industrial Application

The model described above has been used to predict the orientation of a population of fibres in conditions representative of industrial castings. 30 litres of a fibre reinforced concrete are poured in a U shaped channel in three phases. Wall effects are also implemented over 5 mm from the walls for these 10 mm length fibres [38]. Results are shown in Fig. 5.14. Orientation is found to be quasi instantaneous (of order of 1 second) with very high factors close (0.99) to the walls. In the unsheared zone in the central and inner part of the element, fibres stay isotropic, which is consistent with the dimensional approach [39,40].

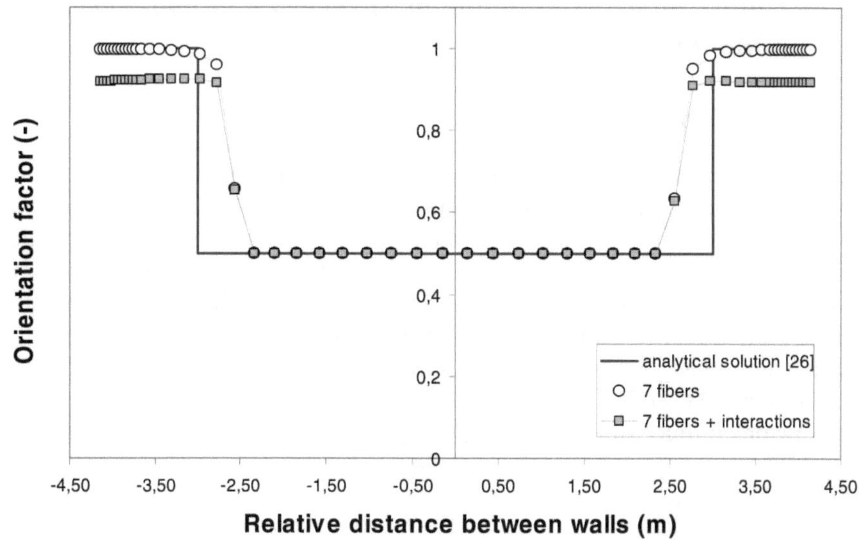

Fig. 5.13 Orientation profile between two parallel infinite walls distant from 20 cm relatively to the flow direction calculated from 7 fibres immersed in a fluid concrete of 300 Pa yield stress and 50 Pa.s plastic viscosity. Comparison with the analytical solution.

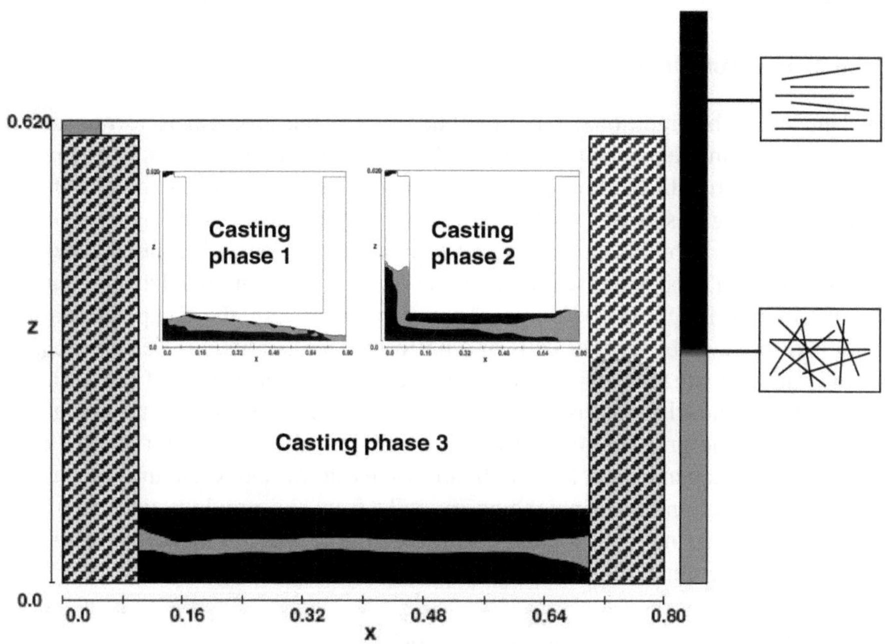

Fig. 5.14 Orientation factor in a U shaped channel (0.6 x 0.2 x 0.1m) relatively to the x direction calculated from 7 fibres immersed in a 300 Pa yield stress and 50 Pa.s plastic viscosity concrete

5.2.6 Fully Coupled Simulation of Suspension of non-Newtonian Fluid and Rigid Particles

5.2.6.1 Modelling Strategy

The complex problem of the flow of suspension of rigid bodies in a non-Newtonian fluid is separated into three levels allowing robust implementation and efficient simulations. The levels, with reference to Fig. 5.15, and their assumptions are: a) Level of particles with exact (analytical) geometry used for dynamics (position and velocities $\underline{\mathbf{u}}$(t), $\underline{\boldsymbol{\omega}}$(t)) and interactions (collisions and force $\underline{\mathbf{i}}$(t)) of particles; b) Level of fluid-particles interaction with particles discretized by Lagrangian nodes $\underline{\mathbf{x}}^L$(t) solved by the Immersed Boundary Method (IBM) with direct forcing $\underline{\mathbf{f}}(\underline{\mathbf{x}}^E,t)$ and c) Level of fluid where the LBM is used for flow of a non-Newtonian fluid with external force $\underline{\mathbf{f}}(\underline{\mathbf{x}}^E,t)$.

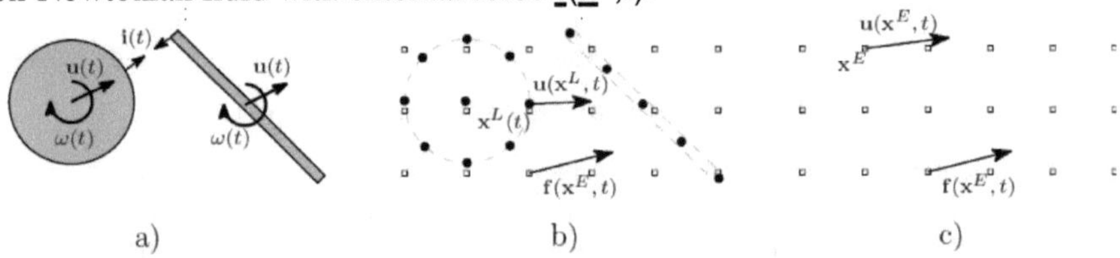

Fig. 5.15 a) level of particles, b) level of fluid-particles interaction and c) level of fluid

5.2.6.2 Level of Fluid: Fluid Dynamics Solver

In contrast to the traditional CFD methods where the problem is formulated in spatially and time dependent velocity and pressure fields (macroscopic quantities, top-bottom approach) the Lattice Boltzmann Method [43], with its roots in the kinetic theory of gases, treats fluid as individual particles discretized by a set of discrete particle distribution functions and provides rules for their mutual collisions and propagation. The macroscopic quantities can be then computed as moments of the distribution functions (bottom-up approach). Typically, the computational domain is discretized by a set of cells of a uniform size. Continuous fields of macroscopic variables are then approximated by sets of average values of the variables in the cells. Similarly, time is discretized into uniform time steps. In a given time step, velocity and pressure of the fluid in a cell described by the particle distribution functions depend on a local collision of particles and streaming of particle distribution functions from neighbouring cells from the previous time step. The local nature of the method, together with the simple algebra involved, leads to a favourably low computational costs and simple implementation.

5.2.6.3 Level of Fluid: Free Surface Algorithm

A free surface has been implemented in the form of a Mass Tracking Algorithm (MTA) [44]. Although the method theoretically conserves the mass exactly, small discrepancies due to the discretization were observed. The difference is negligible in one time step but systematic which leads to its accumulation during the computation making the error important. Therefore, a simple correction based on comparison of actual total mass with its initial value and distribution of the missing mass uniformly to all interface cells is adopted.

5.2.6.4 Level of Fluid-Particles Interaction: Immersed Boundary Method

The IBM with direct forcing [45] represents the particles in the fluid in a form of a force field. It is assumed that the velocity of the particle and fluid should be equal at Lagrangian points due to the no-slip boundary condition. Non-equal velocities are transformed into force field acting both on the particle and on the fluid assuming Newton's second law of motion. Since, generally, Lagrangian nodes do not coincide with Eulerian nodes, the velocity of the fluid in Lagrangian nodes is obtained by a volume averaging of the velocity field obtained from the LBM solution. The IBM, contrary to most methods assuming bounce-back walls at the fluid level, provides smooth and stable change in positions and forces acting on particles and makes *e.g.* coupling with the free surface algorithm easier. However, the most important feature of the IBM lies in its ability to accurately simulate small objects of only a few lattice units or even sub-grid objects, see [46]. This results in a significant reduction of the computational time needed.

5.2.6.5 Level of Particles: Adaptive Sub-stepping Algorithm

The dynamics of immersed particles is driven by the Newton's second law of motion discretized with the explicit forward Euler method. However, when the force acting on the particle varies significantly during one lattice time step (*e.g.* light particles, fibres), the method might become unstable. A variable sub-stepping algorithm has been proposed in [47] to address the problem. In the algorithm the duration of an adaptive sub-step is computed analytically by restricting change in the forces exerted on the. This ensures stability of the simulation for highly variable forces.

5.2.6.6 Level of Particles: Interaction of Particles

Force and direct interactions of particles are solved at the level of particles. The direct interactions (collisions) are simulated using the approach of force impulses. The normal impulse is assumed inelastic while the tangential impulse obeys Coulomb friction. The variable sub-step algorithm ensures stable simulation for arbitrary force interaction between two moving particles. In this contribution a lubrication force correcting term from [48] is used.

5.2.6.7 Application to the Effect of Particles on Effective Rheological Properties

The effect of two types of rigid particles – spheres and fibres – on effective rheological properties was investigated. Suspensions of a Bingham fluid and a varying volume fractions of particles were sheared in the Couette geometry with different shear rates. The effective rheological properties were computed from volumetrically averaged strain rate eliminating the wall effect and stress exerted on the plates. Fig. 5.16 shows the results and their comparison to theoretical solutions. Details can be found in [46-47, 49].

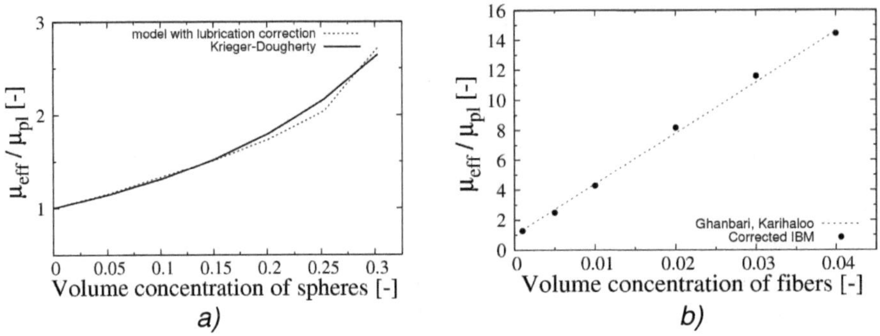

Fig. 5.16 Effect of volume fraction of spherical aggregates (a) and fibres (b) on effective plastic viscosity compared to theoretical predictions

5.2.6.8 Application to Dynamic Segregation in a Complex Flow

A flow of suspension of rigid spheres and a Bingham fluid poured slowly into a rectangular domain was simulated. Fig. 5.17a) shows the configuration of the experiment, see [50] for all details. The final shape of the suspension and the final distribution of the particles and its comparison to experimental findings are shown in Fig. 5.17b). Note that more than 3300 particles were simulated using an ordinary PC.

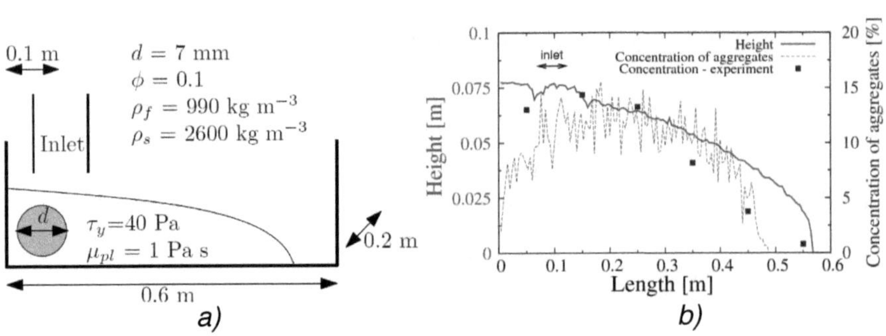

Fig. 5.17 Configuration of the experiment (a) and final shape and distribution of aggregates in comparison with experimental findings (b)

References

[1] Sulsky, D., Schreyer, H.: Antisymmetric form of the material point method with applications to upsetting and Taylor impact problem. Comp. Meth. in Applied Mech. Eng. 139, 409–429 (1996)

[2] Moresi, L., Dufour, F., Mühlhaus, H.B.: A Lagrangian integration point finite element method for large deformation modeling of viscoelastic geomaterials. J. of Comp. Phys. 184(2), 476–497 (2003)

[3] Dufour, F., Pijaudier-Cabot, G.: Numerical modeling of concrete flow: homogeneous approach. Int. J. for Num. And Anal. Meth. in Geomechanics 29(4), 395–416 (2005)

[4] Modigell, M., Vasilic, K., Brameshuber, W., Uebachs, S.: Modelling and Simulation of the Flow Behaviour of Self-Compacting Concrete. In: Schutter de, G., Boel, V. (eds.) 5th International RILEM Symposium on Self-Compacting Concrete, SCC 2007, Ghent, Belgium, September 3-5., vol. 1, pp. S.387–S.392. RILEM Publications, Bagneux (2007) ISBN 978-2-35158-050-9

[5] Modigell, M., Hufschmidt, M., Petera, J.: Two-Phase Simulations as a Development Tool for Thixoforming. Processes Steel Research International 75(9), 513–518 (2004), Larson G.: The structure and rheology of complex fluids. Oxford University Press, New York (1999)

[6] Hufschmidt, M., Modigell, M., Petera, J.: Modelling and simulation of forming processes of metallic suspensions under non-isothermal conditions. J. Non-Newtonian Fluid Mech. 134, 16–26 (2006)

[7] Petera, J.: A new finite element scheme using the Lagrangian framework for simulation of viscoelastic fluid flows. J. Non-Newtonian Fluid Mec. 103, 1–43 (2002)

[8] Martys, N.S.: Study of a dissipative particle dynamics based approach for modeling suspensions. Journal of Rheology 49, 401–424 (2005)

[9] Foss, D.R., Brady, J.F.: Structure, diffusion,and rheology of Brownian suspensions by Stokesian dynamics simulations. J. Fluid Mechanics 407, 167–200 (2000)

[10] Martys, N.S., George, W.L., Chun, B., Lootens, D.: A smoothed particle hydrodynamics based fluid model with a spatially dependent viscosity: Application to flow of a suspension with a non-Newtonian fluid matrix (submitted for publication)

[11] Martys, N.S.: A classical kinetic theory approach to lattice Boltzmann simulations. International Journal of Modern Physics 12, 1169–1178 (2001)

[12] Hoogerbrugge, P.J., Koelman, J.M.V.A.: Simulating microscopic hydrodynamic phenomena with dissipative particle dynamics. Europhysics LettersĨ9, 155–160 (1992)

[13] Landau, L.D., Lifshitz, E.M.: Fluid Mechanic. Pergamon Press, Oxford (1987)

[14] Ferraris, C.: Private communication

[15] Spangenberg, J., Roussel, N., Hattel, J.H., Thorborg, J., Geiker, M.R., Stang, H., Skocek, J.: Prediction of flow induced heterogeneities and their consequences in Self Compacting Concrete (SCC). In: Khayat, K., Feys, D. (eds.) Proceedings of SCC 2010, Montreal, Canada. Springer (2010)

[16] Jeffery, G.B.: Proceeding of Royal Society A 102, 161–179 (1922)

[17] Folgar, F., Tucker, C.F.: J. Reinf. Plast. Comp. 3, 98 (1984)

[18] Rahnama, M., Koch, D.L., Shaqfeh, E.S.G.: Phys. Fluids 7(3), 487 (1995)

[19] Anczurowski, E., Mason, S.G.: J. Colloïd Interface Sci. 23, 522 (1967)

[20] Trevelyan, B.J., Mason, S.G.: J. Colloïd Sci. 6, 354 (1951)

[21] Mason, S.G., Manley, R.S.: Proc. Royal Soc., Series A 238, 117 (1957)

[22] Harris, J.B., Pittman, J.F.T.: J. Colloïd Interface Sci. 50(2), 280 (1975)

[23] Bibbo, M.A., Dinh, S.M., Armstrong, R.C.: J. Rheol. 29(6), 905 (1985)
[24] Kameswara Rao, C.V.S.: Cem. Concr. Res. 9, 685 (1979)
[25] Dinh, S.M., Armstrong, S.M.: J. Rheol. 28(3), 207 (1984)
[26] Bretherton, F.P.: J. Fluid Mechanics 14, 284 (1962)
[27] Goldsmith, H.L., Mason, S.G.: Rheology: Theory and Applications, New-York, vol. 4, ch. 2, pp. 85–250 (1967)
[28] Lipscomb, G.G., Denn, M.M.: J. Non-Newt. Fluid Mech. 26, 297 (1988)
[29] Vincent, M.: PhD-thesis, ENS des Mines de Paris (1984)
[30] Taskernam-Kroser, R., Ziabicki, A.: J. Polymer Sciences 1(6), 491 (1963)
[31] Martinie, L., Rossi, P., Roussel, N.: Cem. Concr. Res. 40, 226 (2010)
[32] Petrich, M.P., Koch, D.L., Cohen, C.: J. Non-Newt. Fluid Mech. 95, 101 (2000)
[33] Kooiman, A.G.: PhD-thesis, DTU, Netherlands (2000)
[34] Ozyurt, N., Mason, T.O., Shah, S.P.: Cem. Concr. Res. 36, 1653 (2006)
[35] Lataste, J.F., Behloul, M., Breysse, D.: Proceedings of the AUGC Symposium, Bordeaux, France (2007)
[36] Boulekbache, B., Hamrat, M., Chemrouk, M., Amziane, S.: EJECE (1985)
[37] Ozyurt, N., Woo, N.Y., Mason, T.O., Shah, S.P.: ACI Materials Journal. Technical Paper 103(5), 340 (2006)
[38] Dupont, D., Vandewalle, L.: Cem. Concr. Comp. 27, 391 (2005)
[39] Martinie, L., Roussel, N.: Simple tools for fiber orientation prediction in industrial practice. Cem. Concr. Res. 41, 993–1000 (2010, 2011)
[40] Martinie, L.: LCPC, Université Paris-Est. PhD thesis (2010) (in French)
[41] Aveston, J., Kelly, A.: J. Mater. Science 8, 352 (1973)
[42] Krenchel, H.: In: Neville, A. (ed.), p. 69. The Construction Press, UK (1975)
[43] Aidun, C.K., Clausen, J.R.: Lattice-Boltzmann Method for Complex Flows. Annual Review of Fluid Mechanics 42(1), 439–472 (2010)
[44] Körner, C., Thies, M., Hofmann, T., Thürey, N., Rüde, N.U.: Lattice Boltzmann Model for Free Surface Flow for Modeling Foaming. Journal of Statistical Physics (2005)
[45] Feng, Z., Michaelides, E.: Proteus: a direct forcing method in the simulations of particulate flows. Journal of Computational Physics 202(1), 20–51 (2005)
[46] Švec, O., Skoček, J., Stang, H., Olesen, J.F., Poulsen, P.N.: Flow simulation of fiber reinforced self compacting concrete using Lattice Boltzmann method. In: Proceedings of 13th International Congress on the Chemistry of Cement, Madrid (2011)
[47] Skoček, J., Švec, O., Spangenberg, J., Stang, H., Geiker, M.R., Roussel, N., Hattel, J.: Modeling of flow of particles in a non-Newtonian fluid using lattice Boltzmann method. In: Proceedings of 13th International Congress on the Chemistry of Cement, Madrid (2011)
[48] Nguyen, N.Q., Ladd, A.: Lubrication corrections for lattice-Boltzmann simulations of particle suspensions. Physical Review E 66(4), 1–12 (2002)
[49] Švec, O., Skoček, J., Stang, H., Olesen, J.F., Poulsen, P.N.: Fully coupled Lattice Boltzmann simulation of fiber reinforced self compacting concrete flow. In: Proceedings of Computer Methods in Mechanics 2011, Warsaw, Poland (2011)
[50] Spangenberg, J., Roussel, N., Hattel, J.H., Stang, H., Skocek, J., Geiker, M.R.: Flow induced particle migration in fresh concrete: Theoretical frame, numerical simulations and experimental results on model fluids. Accepted for publication in Cem. Concr. Res. (2011)

Author Index

MIX
Papier aus verantwortungsvollen Quellen
Paper from responsible sources
FSC® C105338

If you have any concerns about our products,
you can contact us on
ProductSafety@springernature.com

In case Publisher is established outside the EU,
the EU authorized representative is:
Springer Nature Customer Service Center GmbH
Europaplatz 3, 69115 Heidelberg, Germany

Printed by Libri Plureos GmbH
in Hamburg, Germany